Sheep

*The Breeding, Feeding, Shearing and
Management of Ewes and Lambs
– A Classic English Guide*

By A. R. Jenner-Fust

PANTIANOS
CLASSICS

Published by Pantianos Classics

ISBN-13: 978-1-78987-171-5

First published in 1901

Contents

Disclaimer

This book was first published at the beginning of the 20th century. Its contents - medical and otherwise - pertain to this period.

It is reprinted here for historical and entertainment purposes only. The publisher bares no responsibility for any harm or negative events resulting from misuse of the information herein.

Introduction

BEFORE we enter into particulars connected with our subject: Sheep, their Breeding and Management: it would be as well to consider, at some length, the principles on which breeding in general should be conducted. For, the problems connected with breeding, crossing, etc., are both numerous and intricate.

For instance, why does the produce of a Dorking cock and a Cochin-china hen differ so entirely from the produce of a Cochin cock and a Dorking hen? It is so, and, remark, the difference, though more or less varied in quantity, is constant in quality.

Why, again, is the get by a male horse out of a female donkey utterly distinct from the produce of a stallion ass and a mare? So different, indeed, from one another are they, that each has, in England, a distinctive name, the one by a male ass being called a "mule", that by a horse, a "hinny".

As to the original whence our domesticated animals spring, there can be little doubt that, as in the case of wheat and other cereals, of peas and other pulse, they have been fostered and protected by the hand of man, until the rough progenitors of our Devons and Kyloes, have, in the persons of their descendants, become the smooth, finished pictures we see in our exhibitions.

And we must distinguish races from breeds. We may talk of a Devon or a Kyloe as belonging to a race, but the term cannot with propriety be applied to a Shorthorn; a Welsh mountain sheep and a Scotch Blackface, both may be said to be part of a race, but a Leicester and an Oxford-down belong to a breed.

The first person to form the idea of originating a breed of domesticated animals, which should be superior to the native races, the aboriginals, I need hardly say was Robert Bakewell, of Dishley. He began with the sheep, which rough and ragged, small and ill-shaped as was the stock then available, he succeeded by patient selection and considerate matching of parents, in improving into the modern Leicester. With the cattle of his immediate neighbourhood, for he wisely chose the animals nearest to his hand as his materials, he succeeded in producing the "Longhorns", still highly esteemed in the rich pastures round Leicester and Rugby, a breed of cattle that, though some years ago they were getting into disrepute, are once more coming into favour.

Before a man devotes himself to the breeding of any description of stock, he must necessarily form to himself a definite idea of the style, size, and shape of the animal he desires to propagate.

When Thomas Booth, of Warlaby, began improving the cattle of his vicinity in Yorkshire, he had already in his mind the idea of the modern shorthorn. Somewhere about 1790, observing that the valley of the Tees was depastured by a fine roomy stock of cattle, he chose certain of the best examples of them for his parent stock. The defects that he aimed at suppressing were an undue prominence of the shoulder-point, a general soda-water bottle build of the body, too much "day light" under the belly, and a want of uniformity in laying on flesh evenly and firmly all over the frame. Selecting a few cows from the herd of Mr. Broader, of Fairholme (a tenant of the writer's great-grandfather), and putting them to moderate size bulls, Mr. Booth succeeded in laying the foundation of "The Booth blood", many of the most celebrated families of Shorthorns of the present day owing their existence to these Fairholme cows.

The principle upon which these earlier breeders all worked, was the one that is now universally acknowledged to be correct: "Like produces like" an unfailing ground to go upon, and one that admits of a far wider application than is generally allowed, and should be regarded not only in the coupling of the sexes for the propagation of the inferior animals, but also in the continuation of the human species; if more attention were paid to this rule by our heads of families, we should not have so many idiots and scrofulous people about. Nature always avenges an infraction of her laws.

But, while it is perfectly true that like produces like, there is another rule in biology that steps in to teach us caution, and that is the tendency of all animals to "throw back" to some remote ancestor, whose peculiar points, long forgotten probably, suddenly make their appearance in one of his descendants. This is called "atavism" from the Latin word *atavus*, an ancestor, originally, a gt., gt., gt. grandfather, and is frequently observed by the breeders of white pigeons, who, in spite of all their pains to keep their birds pure in colour, constantly find, to their disgust, that, from time to time, black feathers will show themselves in the growing squabs. (Darwin)

Here, then, we find the rule established that it is not enough that immediate parents be of fine shape, good colour, and robust constitution, but they must be descended from families which, for generations, have boasted of desirable qualities, if we are to hope for an offspring that shall not disappoint our expectation.

The form aimed at by all breeders, more or less, of animals intended for the butcher, is that called by mathematicians, the parallelopiped.

A carpenter's pencil gives a good idea of this figure to non-mathematical readers; it is contained by six sides, each of which is a parellelogram. Its proportions are not only beautiful in themselves, but they have a large capacity within small dimensions.

There is no doubt that the earlier breeders of improved stock began their work by coupling males and females within the forbidden degrees of consanguinity. What else could they do? If they went abroad for unrelated males to put to their females, what did they find? Inferior animals, not fit to be named in the same class as their own. Bakewell, after he had succeeded in fixing the type, in search of which he set out, could never be tempted to make use of a strange anim.al, however enticing might be its form or quality: he bred entirely from his own stock.

What said Mr. Booth to the advice of a friend of mine, who had recommended him to introduce foreign blood into his herd? I will, if you will tell me where to find any as good!

Let us now, for a short time, turn our attention to

The Principles of Breeding

Now, the principles of breeding are no longer veiled in mystery-, but, from the constant inquiry to which they have been subjected, and from the very high attainments of those gentlemen by whom those inquiries have been conducted, a flood of light has been thrown on the question, and rules have been laid down for the guidance of breeders which, when faithfully followed out, will invariably prove satisfactory.

It seems, then, that the organs and functions of the animal structure are divisible into three great classes, the *locomotive*, the *vital*, and the *mental*.

The locomotive organs are the bones; the ligaments; the muscles. These are connected with the nerves of motion which arise from the "cerebellum", the back part of the head. The shape, the limbs, and the skin, belong to this class.

The vital organs are the organs of absorption, as the lymphatics; of circulation, as the arteries, veins; the organs of secretion, as the glands. These are connected with the sympathetic nerves, which spring from the "cerebrum", the fore part of the head. The digestive, respiratory, and reproductive organs, with the fat, milk, and other animal products, belong to this class.

The mental organs: the eyes, the ears; the organs of perception, and the organs of volition. The functions of the first are to receive impressions from without; of the second, to perceive, compare, reflect; of the third to will, and, consequently, to throw the muscles into action to fulfil its purpose.

Now, the grand purpose of these inquiries is to determine whether or not one parent, or both indiscriminately, impart their organisation to the offspring. And it is to this point that, in spite of its want of attractiveness, we should turn our earnest attention, for it is owing to the indifference with which it has been so long regarded that breeding has been so uncertain an undertaking.

Men of science, after innumerable experiments, have decided that one class of organs is propagated by the male, and the other by the female. The whole law may be summarised thus; the dam gives the whole of the nutritive organs, and the sire the whole of the locomotive organs. The thinking organs come in equal and distinct portions from both parents.

Following out, then, to its fullest limit this doctrine, we see that, if we desire to improve any part of the locomotive organisation in our stock, we shall look for it in vain from the female; if, on the other hand we seek to improve the nutritive system, we should look for it in vain from the male; that is, in simple terms, if we desire, in cattle for instance, an improvement in the shape, we must look to the bull for it, if we want increased production of milk, or increased tendency to fatten, we must look to the cow. So, in breeding sheep, it is the ram who gives the wool, the ewe who gives the tendency to fatten and the increased production of milk.

When we consider that both parents have a share, a distinct and positive share, in the mental organs, it will not be difficult to see why, after a long and injudicious course of in-and-in breeding, all desire seems to be wanting in the male. Suppose the case of a bull breeding with his daughter and again with his granddaughter. Now, he gives, let us say, the anterior organs to his daughter, thus the two animals become, so far, identical; but, in breeding with his daughter, he may give his posterior organs to his granddaughter; and, as the granddaughter will thus bear both his series of organs — the former from the mother, the latter from himself — it is evident that, as regards these organs, the two are perfectly identical, and the identity of the mental organs destroys all desire for reproduction, the differences which are essential to excitement having no existence.

But, although some of our early improvers were as we have said, obliged, from the nature of the case, to tread a dangerous path, this is no longer the practice of our great breeders. They all have lines of blood, families, of relations it is true, but sufficiently far removed to be matched without fear of the consequences. And it is fortunate for all of us that it is so, for in no other way could the improvement made of late years by crossing have been carried out. The effect of adhering tenaciously to a particular family, or line of blood, has been to confer on the male a peculiarly impressive power, by which his form and substance are transmitted to his offspring, the dam contributing, apparently, nothing towards it. In no breed is this so marked a quality as in the Shorthorn, probably because in no other case has so much pains been taken to preserve the lines of blood pure and intact. The writer saw, at the Hon. M. H. Cochrane's, a few years ago, a calf, by Royal Commander out of a Kyloe heifer, that disowned, in everything but the jolly ruggedness of his roan coat, the very mother that bore him! Hence it is that four crosses of Shorthorn blood are considered enough to admit the produce to Herd-book honours.

Crossing

Nobody can doubt about the wonderful good that has been produced by the well-conceived system of crossing that has now, for many years, been practised in England. At first, the principle upon which the practice was based was little understood, but of late years the more abrupt and violent attempts of the earlier breeders have been avoided, and the more natural, and therefore more sensible course has been pursued. One rule, however, has been in vogue for, I can safely attest, the last 60 years, viz., to put the best of the pair, i.e. the male *atop*, and to employ in crossing, as in every other branch of breeding, nothing but thoroughbred males.

Before beginning to improve our stock of any kind by crossing, it is evident that we should put this query to ourselves: is our soil sufficiently good to feed the product of the purposed cross as it will need to be fed? There are many situations in which a high-bred stock of sheep, cattle, or horses cannot be maintained as a breeding-stock; continued crossing cannot in such cases be followed out, for, eventually the whole herd or flock would become like the thoroughbred parent, and utterly unfit for the locality. Thus, to keep on breeding from Shorthorn bulls and the common cow of the country, — what the Americans call "scrubs" — on the poorer clay and sands of this province, would be most injudicious. The first and second crosses are all that should be attempted, the heifers being still kept true to the parent stock, until the general improvement, to which we all so anxiously look forward is secured, and the land becomes fit to support a superior class of animals.

As for the notion that exists, that, if a large sire be put to a small dam, the foetus will be so large that the mother will not be able to bring it to the light, I attach no weight to it; for the foetus is always proportioned to the matrix that contains it. There may be a little trouble, perhaps, in its expulsion on account of the increased size of the brain of the improved offspring. I have bred, in cattle, sheep and horses, from all sorts of males, and never found any difficulty on this score, but I have found that where a thoroughbred male of good form has been put to a rough country-bred mare, the progeny was invariably far superior to the dam in all *outward* parts, and that the rugged healthiness of the dam, with her abundant flow of milk gave her plenty of strength to bring forth and to sustain her better bred foal afterwards.

Look, for instance, at the modern Exmoon ponies. Seventy or eighty years ago, they were little creatures from ten to twelve hands high, with nothing but their constitution and hardiness to recommend them. Now, crossed as they have been with full-sized, thoroughbred stallions, a more perfect type of pony for a lady's phaeton cannot be found, and they have so increased in height that many of them run from 14 to 14½ hands.

Again; I have used Shorthorn bulls of such size and weight on small country cows that the latter could barely support the weight of the bull at service, and the parturition of the calf was never attended by any evil consequence more than is usual in an ordinary herd.

Once more; I have coupled the small refined type of Southdown ewes with the heavy Hampshiredown ram, and, although the head of the Hampshiredown was in those days disproportionately large, the lambing was got through with much as usual.

And here I must mention, as an instance of the effect of crossing, the creation of the Babraham flock of Southdowns, premising that what I am about to state is merely an opinion, I formed purely from my own judgment, without the slightest evidence, material or otherwise, to go upon.

The original progenitors of this most beautiful collection of sheep were brought, somewhere about 65 years ago, from Glynde, Sussex, the well known abode of the celebrated Mr. Elhnan, one of the first great breeders of the Southdown. They were elegant, deer-like creatures, with narrow chests, light forequarters, good, though of course small, legs of mutton, and good loins. They "went to fold" every night on the fallows, on the breezy downs looking over the sea. We shall hear more about this folding system later. Seldom killed before three years old, their weight, when fat, varied from 14 to 16 pounds the quarter. To this very day, the highest prices quoted in the papers for mutton in the great London market is for Southdowns; from 60 to 64 lbs. carcass, i.e., 15 to 16 lbs the quarter. What the flavour of the meat of these small sheep was, those who have been so fortunate as to taste the saddle of a three or a four year old Southdown wether will recall with pleasure: such meat is not to be met with nowadays, as it would not pay to keep sheep to that age.

From this flock of Mr. EUman, Mr. Jonas Webb selected a few ewes and a ram which he took with him to a small farm in Cambridgeshire, belonging to Mr. Adeane, of Babraham, whose game-keeper, Mr. Webb was at that time. What the subsequent treatment of the flock was, nobody, I believe, ever knew, but it was most successful; as the wethers at 20 months old often weighed from 22 to 26 pounds a quarter, and yet retained all the quality of the parent stock as regards meat and wool, while the bosom was enlarged, the neck strengthened, and the wool much improved. Of course, the better keep of the Cambridgeshire farm must have had an immense influence on the growth of the animal, for no doubt the small size of the Sussex sheep in its native county is, in great measure, due to the absurd practice of the flock-masters of that county of sending the ewe-lambs out into the poor farms of the lowlands (*the Weald*) at so much a score for the winter, to *harden* them, as it is called! And, truly, it ought to harden them, for when we saw the *tegs*, as the year-old lambs, are locally termed — *hogs* or *hoggets*, in other parts, — just before shearing time, they had returned from their winter quarters mere bags of bones. I remember seeing a lot, from the flocks of three of the best breeders in Sussex, Mr. Tanner, who had then 2,000 ewes, among them, and it was difficult to believe that sensible men could expose their young stock-ewes to such hardships. It must be remembered that the Sussex flock-master sold all his wether lambs at the autumn fairs, retaining at home only the older ewes.

Well, perhaps I was mistaken about the cross of the long-wool in the Babraham flock, and at all events it was a most masterly combination. On one fine day in July, 1852, I passed a couple of days at Babraham and saw the *annual sale* and letting of the rams and lamb-rams. How many were let and sold, I do not recollect, but I have kept the amount fresh in my memory:

£3,752, or $18,760! So carefully had the flock been bred, that the following year I saw 150 rams, just shorn for the first time, turned out of a large shed, from among whom it would have puzzled the best judge in England to select the most perfect, so equal were they in build, colour, and carriage.

Another successful instance of crossing is the well-known Oxford-down, now, more properly called Oxfords. It was only after a 20 years endeavour to form a permanent type of sheep, by the union of the Hampshire-down and the Cotswold, that Samuel Druce, of Eynsham, Oxfordshire, succeeded. At first, and for a long time, the legs were some dark, others light; the faces of some were white, of others brown, and of others mottled; there was no uniformity anywhere. Now, however, all this is changed; the type, or character, has long been fixed, and the sheep in one of the leading ram-breeding flocks are as uniform in character and colour as the sheep in a flock of Leicesters or Cheviots.

In crossing, we should aim at combining utility and beauty, though this union is almost a necessary sequel of judicious selection, for

> "Beauty never deigns to dwell
> Where use and aptitude are strangers."

All attempts at crossing should be kept within certain limits, and a clear idea formed, before beginning, of the object in view, and this idea, when once formed, should be firmly adhered to.

For instance; the cross of the Cotswold ram and the Hampshire-down ewe, as we have seen, turned out most successfully; but the cross between the Cheviot ram and the Leicester ewe, as well as that between the Black-face ram and the Leicester ewe, were complete failures; the progeny, in both cases, showing a worse and more uncertain organisation than either of the parents.

If it be true that breeding from a good sire and dam does not necessarily ensure a good progeny, can it be true that "Like begets like?" I answer, that I believe the adage to be true in a certain sense, but it is undoubtedly not true in the popular sense in which it is used, and I believe it has led many a young breeder astray, by inducing him to believe that, when he had purchased a good-looking sire, he had secured all the conditions necessary for a good progeny. There is no error more prevalent among young breeders and there can scarcely be a more fatal one. An animal has certain qualities apparent to the hand and eye; it has also hidden qualities that neither the hand nor the eye can detect, but which hidden or latent qualities descend to the offspring, and, when the animal has been crossed with another animal of different blood, these qualities will produce new combinations palpable and unexpected. The above maxim is true then in this sense; that although the offspring may appear unlike either father or mother, yet the peculiar properties of both parents are not lost in the offspring; they are inherited, but in combination may have produced effects that probably had not, and could not, with any degree of certainty, have been foreseen. That these qualities are not lost would appear evident, as it is found that peculiarities derived from even remote ancestors will, from time to time, more or. less frequently, according to the skill and perseverance of the breeder, show themselves, or "crop out," to use the term of the geologist.

The law of crossing is that when each parent is of a different breed, both being equal in age and vigour, the male gives the back head and locomotive organs, the female giving the face and the nutritive organs. This law, in its effects on the domesticated animals, is very similar to the law of selection; but in crossing, the parents always maintain this relatively position, while in ordinary breeding, the parents change position in proportion to the comparatively greater vigor of the characteristics of each, and when one imprints the prevailing characteristics, the other stamps the opposite.

The cause that, in the crosses, the male gives the *cerebellum*, or back of the head, and the locomotive system, is both striking and beautiful. If no being can desire that of which it is already in possession; if, on the contrary, it must desire that of which it is most devoid (if not incompatible), it cannot be wondered at that in crosses, when the desired difference is greatest, the male, whose desire is more ardent, should stamp the system by which he exercises that desire, namely the voluntary locomotive, upon the progeny. If, then, of the two great series of organs described, each belongs entirely to a distinct parent, we can neither desire in the progeny both series from one parent, nor portions of both from each parent; and every attempt to attain either end must be a failure. It moreover shows that, in a feeble or imperfect cross, bad as well as good combinations may be produced; but that such a progeny has present the desired qualities must be alone preserved for future breeding, while the inferior must be set aside. The intermediate character of the qualities produced in crossing is owing, not to each of the parents imperfectly contributing its share in the progeny's organization, but that, in their new combination, each series of organs acts upon and therefor modifies the other. [1]

[1] Walker: "On intermarriage.

Chapter One - Of the Different Breeds of Sheep

SHEEP, in general, may be divided into two classes: long-woolled and short-woolled; though occasionally it may be convenient to speak of the middle-woolled, such as the Oxfords, Shropshires, etc.

All the Down-breeds belong to the short-woolled class; all the white-faced, such as the Lincolns, Leicesters, etc, to the long-woolled. The former are superior in the flavour of the meat, and more active in their habits, than the long-wools; they stand *crowding* better, *i.e.*, more can be kept on the same extent of land, and their wool, in general, fetches a higher price on the market.

In this country, particularly in this province, owing to the prevailing practice of domestic manufacturing, it has become the habit of the farmer to look upon the sheep as a wool-grower rather than as a producer of meat for the table; consequently, it any improvement was sought for in the older flocks of the French-Canadian sheep, it was found in the long-wools, the Leicester or the Cotswold; and this may account in great measure for the inferiority of the greater part of the mutton we meet with at the hotels in the country towns of the province. Before the day of the late lamented Major Campbell, of St. Hilaire, there was hardly a Down sheep to be met with in the country; thus, those farmers, who really would have improved their flocks if they could have found the where-with, were left stranded, so to speak.

However, in spite of our undisguised partiality for the short-wools, we must not forget that the great majority of the sheep in the country are still long-wools, and that a brief description of the different breeds will be found interesting to their owners.

Chief breeds of long-woolled sheep kept in the province of Quebec are the Leicester and the Cotswold, with a few small flocks of Dorset-horns, that should rather be reckoned among the middle-wools.

The Leicester

The Leicester, so called from the well-known county of that name, is one of the earliest instances of what can be done by the improver's skill and energy. Robert Bakewell, of whom more hereafter, was the man who first, so to speak, *made* this valuable breed of sheep. The wool of the Leicester is not so long as the wool of the Lincoln, but, there is no doubt of its having been employed in the remoulding of the other long-wools. As far back as 1668, Markham, writing of the sheep of the English Midland counties, speaks of a large-boned sheep, of the best shape, and the deepest staple (the thread or pile of the wool); they were chiefly pasture sheep, and their wool was coarser than that of the Cotsal.

Leicester Ram

Note. — We may remark here that "Cotsal," or in the modern form, "Cotswold," a range of hills in Gloucestershire, is not, as usually supposed, derived from cot, a mud-hut, and wold, a wood, but from cote, a wood, and weald, a cleared forest, a curious example of two synonymous elements meeting in the same word, which often occurs in English names of places.

Professor Low also says: "There is no reason, therefore, to assume, from any of the characters presented by the wool of the new Leicester breed, that the parent stock was any other than the long-woolled sheep of the Midland counties."

It was thus from the ordinary sheep of his district that Robert Bakewell, a man of peculiar character, by dint of thought and determined perseverance created the New Leicester or, as they were commonly called at first, the Dishley breed. His success was due to a firm faith in the power of animals to transmit their good qualities to their progeny, and to his constantly keeping his eye on the type he wished to produce. Apparently, he did not greatly trouble himself about the wool, but aimed at bringing out an animal taking in form, of early maturity, and possessed of a tendency to lay flesh on the upper part of the body.

It was about the year 1755 that Bakewell began the improvement of the breed of sheep that lay at his own door. He worked upon the principle, of which he was the originator, that the properties of the parents may be transmitted to their offspring until fixity of type is the result. He had also what in modern slang would be called "a good eye for a beast," whether that beast was a sheep, a bull, or a stallion. He could see that quality was preferable to mere bulk; that perfection of form must be accompanied by readiness to profit by food.

The result of Bakewell's work was the formation of an improved sheep, somewhat smaller than the older breed, but more symmetrical, thicker, deeper, and more easily fattened at an earlier age. He reaped a great harvest from his exertions; his first ram let for, in our currency, $4.50 the season, but before long the price rose to $500.00, and in 1786 he made 15,000.00 by letting his stock. In 1789, he let three rams for $6,000.00, and $10,000.00 for seven; and in the same year he made $15,000.00 more by letting the remainder of the rams to the Dishley Society, then just founded for the purpose of carrying on the good work.

The above facts, says Mr. Wrightson, must appear extraordinary to any one who reflects upon the greater value of money one hundred years ago than now, and the much less general appreciation at that time of the advantages of well-bred stock. Then, there were no foreign buyers to stimulate biddings, no princes or millionaires competing for favourite strains of blood.

The Improved Leicester

The modern Leicester is a white-faced, hornless sheep, covered with a fleece, the staple of which is from seven to eight inches long, with a short twisted curl at the end.

Points: — Nostrils and lips black, nose rather narrow with a tendency towards the Roman, but the shape of the face in general like a wedge, and covered with wool as to the forehead, though I have seen many good sheep with a naked front; no signs of a horn: ears thin, long, and mobile; a black spot occasionally on face and ears; a good eye; neck short, and level with the back, thick, and tapering from skull to shoulders and bosom; breast deep, wide, and prominent; shoulders somewhat upright and wide over the tops; great thickness through the heart; well filled up behind the shoulders, giving a great girth; ribs well sprung from the back-bone, loins wide, hips level, quarters long and straight, well set-on tail, good legs of mutton well rounded over the hock, barrel round, great depth of carcase, fine in the bone; the fleece curly and free from black hairs; the back and loins well-covered, the flesh firm, the pelt or skin springy to the touch; the legs well set-on, hocks straight, pasterns firm, and feet neat.

There is no use in keeping Leicesters too long before fattening them for the butcher, for they get so fat after they reach the age of fifteen months, or so, that, as somebody says somewhere, no one but a Scotch collier would eat them, and thereby hangs a tale: Many years ago, in 1848 or 49 I think, I was passing a few days at the Royal Hotel at Norwich, in a county where the farmers were all extensive sheep-men. Fancying that, in such a town, the mutton must be good, I ordered, for the dinner of myself and party, a leg of mutton, among other things. At the appointed hour, up came the mutton; I carved it, helped my friends round, and then myself. "What on earth is this?" exclaimed I, on tasting it. "The mutton you ordered. Sir," replied the waiter. "Oh!" said one of my guests, who happened to be a resident of the county; most of our sheep here are Leicesters; perhaps long-wools are not kept where you come from, "No, said I, and if this a leg of long-wool mutton, I hope they never will be, for it is the worst flavoured mutton I ever tasted." Without stating that I had never eaten anything but Down mutton before, I may safely say that the flavour of the Norwich meat was so strange to me that I thought some one had been playing me a trick to see if I should find it out.

Says Prof. Wrightson, "There is an idea prevalent that the day of the Leicester is gone by. Purebred Leicesters are so much given to la}' on fat thickly," (and that on the loin and neck especially) " and the demand for fat meat of all kind has so completely ceased, that everywhere the Downs are preferred." The great value of the Leicester is for crossing, and, before I left England, the flocks in the county of Norfolk, where, as I said before, the Leicester reigned, in 48, pre-eminently, had almost all been converted into half-breds, so it would be wrong to say that the Leicester is played out. A useful cross is that between the South-down and the Leicester, or the one between the Hampshiredown and the Leicester. Of the numerous crosses used between the Leicester up in the North-country I can say nothing, for I am not familiar with them, but there is, according to those who know what the}' are talking about, scarcely a breed which has not felt the influence of the Leicester, from the Cheviot and the Black-face of the Borders to the Cotswold of the hills and the Lincoln of the fens.

Lincoln Ram

Lincoln

As we have just mentioned the *Lincolns*, we may as well devote a few sentences to them, though we do not fancy there are many of them kept Canada.

The Lincoln is not unlike a large Leicester. It is the heaviest breed of sheep known, a full-mouthed ram of that breed having been killed, in 1826, that weighed 96½ lbs. the quarter! The flesh handles more firmly than the flesh of the Leicester, the wool is extraordinarily long, samples having been met with that measured 21 inches in length; the whole fleece of the above-mentioned ram weighed 30 lbs.

By the bye, it would be well to say that whenever the weight of a fleece is stated in this essay, it is to be understood that it is the weight of a fleece that has been thoroughly washed on the sheep's back before shearing. I shall describe the operation of washing in the chapter on shearing.

The wool of the Lincoln is very bright and lustrous; hence it has gained the name of lustrewool, though it loses that character when taken away from its native habitat. In 1788, Bakewell got into a famous dispute with Mr. Chaplin, an ancestor of the member of Parliament who, not very long ago, was Minister of Agriculture in Lord Salisbury's cabinet. It seems that Bakewell was supposed to have been prying about Chaplin's rams, after having been refused leave to inspect them, and Chaplin, naturally, was not pleased. However, Bakewell got out of the scrape pretty easily, showing that he had been introduced to the flock by Chaplin's own man, and had not, as Chaplin accused him of doing, "been meanly sneaking into my pastures at Wrangle."

The new or improved Lincoln is the product of lyicester crosses upon the old Lincoln, and according to my idea can only be seen in perfection in his own home, as he needs, to what M. Mirobolant calls "perfectionate his work," rich pastures and lots of room. As a mutton sheep, he is inferior to the Downs, as far as quality goes, though you must not tell a Lincolnshire man so, for I recollect some years ago a young Lincolnshire farmer, *at Sorel*, who was very anxious to send, on his return to England, a few Lincoln ewes and a ram to begin a flock of that breed in that notoriously poor sandy spot!

The highest prices paid for rams of late or, indeed, at any time, have been paid for Lincolns, a ram of the breed having been sold for a thousand guineas, $5,000.00, only last year! He was bred, if my memory serves me, by Mr. Dudding of Riby.

Border-Leicester

I have not heard of late of any *Border-Leicesters* being kept in this province, but I remember a small flock of them, in 1870, being on the hands of the late Mr. Thomas Irving, of Petite Cote, on the island of Montreal. If I do not mistake, the late Judge Ramsay had some of them, for I remember a long-legged brute of a ram, that could jump like a well-bred hunter, and who knocked me over (from behind!) when I was at the Judge's place at St-Hugues. The breed, however, was in high favour in the counties of Cumberland and Northumberland, and came out well at the last meeting at Windsor of the Royal Agricultural Society of England.

Border Leicester Ewe

The Border-Leicesters owe their improved state to the Culleys, who farmed on an immense scale at Wooler, Northumberland, paying rent for their several farms to the amount of £6,000 or $30,000.00! At that time, 1767, the long-legged, rough-woolled *Tees-Water* was the chief breed of that county, but the new Leicesters, brought from Dishley by the Culleys, soon made a change in the appearance and quality of the flocks. In 1888, Ivord Polwarth sold from the Mertoun flock 28 shearling rams for an average of £36. 9s. 3d., and one for 165 guineas, or $182.00 and $825.00; in 1890, the Mertoun rams averaged £53, 19s 4d., or $268.00.

Cotswold

Unlike most of the long-wools, the Cotswold seems to have early taken a fancy to the bleakest range of hills in the West-midlands of England. Rising from the River Severn, the lovely vale of Glo'ster spreads its rich meadows along the flat, gradually sloping upwards, till after terrace upon terrace has been mounted, the

18

view is arrested by the steep, abrupt escartpment of the Cotwolds. Cold and bleak from their exposed situation, the hills produce rare food for stock. The pastures, though not what our fathers would have poetically termed "lush," afford excessively healthy grass, and the rootcrops are both abundant and nutritious.

The Cotswold sheep are among the most ancient of our reorganised breeds. "Here" says Camden, "they feed in great numbers flockes of sheep, long necked, and square of bulk and bone by reason, as is commonly thought, of the weally and hilly situation of their pasturage, whose wool being more fine and soft is held in great account by all nations."

Stowe, in his Chronicles, says that, in 1463, Edward IV "concluded an amnesty and league with King Henry of Castille and King John of Aragon, at the concluding whereof he granted licence for certain Coteswold sheep to be transported into the country of Spain, which have there since mightily increased and multiplied to the Spanish profit."

But were these fine-wooled sheep the Cotswolds as we know them? Hardly, I think; they were probably more like the merino; for the pasture on the hills was shorter and finer than the grass on the rich grazing lands of Leicestershire and the other "Shires," as our hunting men *Quorum pars parva qui* term them, and short, fine grass would naturally produce short, fine wool.

There is no doubt that in the 18th century many Leicester rams were imported into the Cotswold country, and the breaking up of the downs, followed by the growing of turnips, would tend to increase the size of the sheep and the length and strength of their wool.

All Canadians who care about sheep, know a Cotswold when they see one; but many Canadian breeders of Cotswolds do not know that a gray-face is by no means a disqualification. I constantly see in the *Gloucester Chronicle*, a paper I receive every week, advertisements of flocks for sale by auction, in which *gray-faces* are mentioned as enviable characteristics of the sheep in question.

The Cotswold is a big, upstanding sheep, with a more *dégagé* carriage than the I,incoln or Leicester. His heavily woolled forehead is his great distinguishing mark, though his wool in general, with its bold, open curl should, to any judge of sheep, at once differentiate him from the other long-wools. Rather, "ewe-necked" and "goose-rumped" it is true, but, on the whole a good "body of mutton;" his shoulders broad on the tops, legs well let down, ample loins and well-sprung ribs; all these points show him to be a really valuable sheep for every table except where small joints are in request, as in the case of the *West-end* of London, to which market, it is to be hoped, all our Quebec mutton will find its way when our farmers learn their true interests and devote themselves to the production of the best breeds of both cattle and sheep.

The fleece of the Cotswold runs from 6 to 8lbs., though some exceptional fleeces may be met with that go over 8 and even 9lbs.

Though two or three years ago, at the Smithfield Club show, in London, a pen of Cotswold lambs was nearly, if not quite, at the head of the lambs exhibited the Cotswold, as a general rule, is by no means celebrated for early maturity. They are said, by those who know them well, to be delicate when young, and to require time to come to the knife. They will not bear the close folding that suits the Down breeds, but have to be kept in small lots, a fact that struck us forcibly when we first ran against them, unexpectedly, at a well-known farm near Compton, some 28 years ago.

"Cotswold mutton," says Prof. Wrightson, is of second quality, like most of the long-woolled breeds, and is pale in colour and long in the grain. When long-wool was higher in price than short-wool, the position of the Cotswold was stronger than at present as a Cotswold fleece was well worth a sovereign — $5.00 — . Now, the same fleece is probably not worth more than ten shillings. For a time, the demand for Cotswold rams seemed in danger of falling off, but during the last two or three seasons there has been a reaction in their favour.

Chapter Two - Short Woolled Sheep

Not much use in talking about the Kents, or the Wensleydale breeds of sheep, as there are none of either in this country, so we will plunge at once into the study of the great exponents of the true short-wools, the Down breeds; and first of the true protagonist of the clan, the

South-Down

The true habitat of this charming sheep is the Downs of the county of Sussex. Having been a pupil of the late Wm. Rigden, of Hove, near Brighton, to whom I went for six months for the express purpose of studying the Southdown "at home," and having found him always ready to impart any information of which he was possessed, I may say, I think without vanity, that I know what a South-down sheep is as well as any man not born among them, can know. The pleasantest six months I ever passed in my

South-Down Ram

life were passed under Mr. Rigden's roof, and our daily visits to the ewes — for I fortunately arrived at Hove just as the ewes were lambing, — were full of interest.

A very curious thing struck me almost the first day: the shepherd knew every ewe by sight, could name their sire and dam at a glance; but that was a mere trifle, compared with his skill in detecting any ewe with a tendency to neglect her newly born offspring; in pouncing upon her briskly, though gently and compelling her to discharge the duties of her position.

Mr. Rigden often used to laugh at his first attempt to win honours at the R. A. Society's show of breeding stock. It was, I think, at Manchester he first exhibited, and with what glee he used to recite the speech of one of the brothers Webb who, looking at Mr. Rigden's best Southdown ram, cruelly told him that "he had better tie it round his neck and give it a dip in the sea at Hove! This was some time about 1843, or '44, and within five years of that time, Mr. Rigden was winning prize after prize all over the South of England, and at last, when the Grand Monarque, Jonas Webb, gave up exhibiting, Rigden reigned, *facile princeps*, over all the Southdown breeders, and as Prof. Wrightson says: "The Southdown race in its own district was long well maintained by the late Mr. Rigden," though that is but tame praise from such a pen.

Have my readers any idea of what the Downs in the South of England are like? Stretching from the *weald* (same as *wold* in Lincolnshire, and *wald* in German) runs a series of low hills, succeeding one another, like what the Ontario people call "Rolling-land", only more acute. These hills, downs (in French *dunes*) are all situated on the chalk rock, which acts as an all-pervading drain, so that within an hour or two after the heaviest rain the surface is as dry as it was before the rain fell. Owing to this cool subsoil, the soft, nutritious herbage, though always short, never withers, and the closer it is fed off the denser and more succulent is the pasture.

Naturally, one would not expect to find animals with broad briskets and large frames on such land as this, and, truly, the sheep of the South-downs are not large, in spite of the extra good feeding they receive as compared with the semi-starvation they used to be obliged to endure in the early part of the last century. Even now, the ordinary Sussex Down is small, and were it not for the high prices the wealthy people of the West-end of London and the visitors to the watering-places of the south coast are willing to pay for, as I said before, small joints, the breeding of this style of sheep could never pay the farmer. The favourite weight of the best sheep in the London market is from 7½ to 8 stones, of 8 lbs. to the stone. You will see by the subjoined list of prices at the Islington cattle market how much size regulates the value of sheep there:

7 " to 8 stone Downs	$1.50
8 " Scotch	1.50
9 " Downs	1.46
10 " Irish	1.28
10 " Half-breds	1.35
10 " Downs	1.39
10 " Ewes	1.07
12 " Half-breds	1.24 to $1.26

So you see plainly that in proportion to the rise in weight, down go the prices, for lo stone ewes are worth five cents a pound less than 8 stone Down wethers.

The Scotch sheep mentioned in the above list are the four-year-old black-faced wethers, and barring their want of fat, very fine mutton they are, though I, probably from being brought up on it, prefer the Down meat.

Ellman of Glynde is still "a name to conjure by" in all the Down country, for it is to his earnest perseverance and skill that the old-fashioned Sussex Down, small in size, bad in shape, long in the neck, low at both ends, full of faults everywhere, except in the leg, was first improved.

The Southdown is unquestionably the fashionable sheep; George III having taken it up, plenty of "his nobility" followed suit, and to-day many of the leading flocks of this breed belong to such men as the Prince of Wales, [1] the Duke of Richmond, Lord Walsingham, &c.

Southdown mutton is close-grained, dark in colour, tender and juicy, and the proportion of lean to fat is just what it ought to be. To use the butcher's phrase, "it dies well," for there is always plenty of internal fat and such.

No doubt this breed originally wore horns, but, by dint of selection, these have been "bred-out," except that, here and there, slight lumps or slugs, as the shepherds call them, make their appearance: a clear case of "heredity." I remember well, when I was young in the business, the well-known John Clayden, of Littlebury, pointing these slugs out to me in the forehead of a ram I had hired of Jonas Webb.

As for the light forequarter, as Professor Wrightson observes, what else can be expected in a sheep the muscular development of whose hind-quarters so necessary in an animal that had to climb such steep hills, necessarily implies the contraction of the fore-end.

The Southdown seems to have an especial partiality to chalk soils and not too rich keep; at least it soon alters both in form and wool when taken to other climes and other pastures. I remember well Lord Ducie, the fortunate owner of the "Seventh Duke of York" and other valuable bulls of the "Bates blood," hiring a Southdown ram from Jonas Webb for 100 guineas the season. He kept the same ram for three seasons, but the progeny of that ram was no more like the sire than I am to Hercules. The form of the lambs was loose and straggling, the head coarse, and the wool opened and what used to be called "lashy." An utter failure, owing entirely to the change from the short, scanty pasture of the Downs with their free and bracing air, being exchanged for the close, confined climate of the Vale of Berkeley and the Irish herbage of the Tortworth meadows.

[1] Written in 1900.

Chapter Three - The Hampshire-Down

WE have now arrived at my own favourite breed: I am not going to pretend, as some have pretended, that this breed of sheep is absolutely free from all admixture of blood. It is a Down, emphatically, and, no doubt, its race has been improved by the selection of rams from the more refined tribes of the more Eastern hills.

And if I am asked why I have so high an opinion of the qualities of the Hampshire-Downs as being the very sheep for this country, I would simply ask my readers to look over the following article, written for the "Journal of Agriculture", some years ago, but expressing my views of to-day as it did my views in 1883.

Royal 1st Prize Hampshire-Down Shearling-ewes (Hillhurst flock.)

I may as well say that, when farming in England, I kept a flock of 250 ewes, all but 20 of which were pure-bred Hampshire-Downs, the remainder South-Downs from Mr. Rigden's flock. The Hampshire-Downs were not fancy-bred sheep, but regular farmer's sheep bought at the fairs in the county whence they take their name.

Hampshire-Downs at Islington, Eng.

At the Christmas Show of the Smithfield Club 1882, these sheep again made their mark. The question of their superiority as regards early maturity may now be considered as definitively settled. It is very much to be regretted that no man, no body of men, has shown sufficient interest in the welfare of the agricultural population of the province to import a few of these most useful sheep. The price is not out of the way: a *good* ram lamb can be bought for £10, and ewes would cost about £4. 10 a piece. Not show sheep; but honest farmers' stock. The ram should, of course, be selected from a family not too nearly related to the ewes.

The following is an analysis of the live weights of the lambs exhibited at Islington; three to each pen:

Cotswolds	595 lbs.
Leicester	558
Lincoln	616
South-Downs	525
Shropshires	451
Oxfords	460
Hampshire Downs	672

And from this list, I deduce the following most important facts: That the Hampshire-Down lambs were superior in weight to all the others, and not by a trifling amount either, as the next table will show:

Weight of Hampshire-Down lambs

672 lbs	=	weight	of	Cotswold lambs	+	77 lbs.
"	=	"	"	Leicester "	+	114 "
"	=	"	"	Lincoln "	+	56 "
"	=	"	"	Southdown "	+	147 "
"	=	"	"	Shropshires "	+	221 "
"	=	"	"	Oxfords "	+	212 "

And more; while the Southdown *wethers* weighed 682 lbs, the Hampshire-Down *lambs* weighed 672 lbs; the former having only 10 ft)s per pen of three, or 3 1/3 lbs each, to show for their twelve month's food! Again; we see by the table *two Hampshire-Down lambs* weighed as much as *three Shropshires*, and nearly as much as *three Oxfords!*

Lastly, the pen of three Hampshire-Down *lambs* exceeded in weight the pen of three Southdown *ewes* (3 years old) by 56 lbs!!! The difference between the weight of the Hampshire-Downs and the Southdowns I was prepared to see, bat I must confess I was astonished at the amazing superiority of the former over the Shropshires and the Oxfords. Judging from my own past experience of sheep in the state of fatness in which they made their appearance at the Smithfield Club Show, I believe I am not wrong in taking 65% of live weight as the weight of the four quarters; which would make their value in the London market, at present prices, £7.6 sterling, or $36.00 i.e. 1882! Most of my readers know, by this time, that in the English markets all cattle, &c. are sold by hand, and the price of mutton is so high now in that country, that a good Down wether weighing, when slaughtered, 20 lbs a quarter, is worth one shilling sterling a pound, or $20, as he stands. I have no hesitation in saying that as long as prices keep up to what they are now, no more profitable system of farming can be offered to the Canadian than the breeding and fattening of sheep for exportation, if the sort of sheep suited to the trade of the *west-end* of London be selected. Hampshire-Downs lambed in March, and moderately pushed from weaning, should weigh, by the time the first boat leaves for England in the Spring, something like 12 stones, or 96 lbs, and would bring in the neighbourhood of twenty-four dollars, and there is only one secret in their management in this country: *rape, rape, rape,* from the 20th of June to the end of the season. It would add at least one-third to the yield of our farms.

Hampshire Down Ram

Well, now that the Hon. Mr. Cochrane and his son have been good enough, — may I say, owing to my repeated prayers? to import a fair number of ewes, lambs, and rams from the best flocks in England, I think there is a fair prospect of my special favourite being, before long, widely known throughout the province; and they only require to be known to be appreciated.

As for the original old West-country sheep, whence the Hampshire-Down derive one side at least of their origin, we know pretty well what they were like. Some of them had horns; all were more or less ragged in appearance; their legs were long and carcase narrow, and the faces and legs white. "These sheep", says Youatt, "not only prevailed on the Wiltshire Downs, and were much and deservedly valued there, but were found in considerable numbers in North-Devon, Somersetshire, Berkshire, and Buckinghamshire. If they were rather slow in fattening, they were excellent folding sheep, and enabled more grain to be grown in Wiltshire than in any other county in England. In

1837, these Wiltshires have all passed away." They were crossed again and again with the Southdowns' until every trace of the old breed disappeared, and a useful variety of the Downs remain a very fine flock of which I remember to have seen at Wenvoe Castle, Glamorganshire, the family seat of one of the Jenners.

At the Oxford show of the Royal, 1840, these modified sheep were shown as "West-country Downs," and are reported to have been not unlike the present type, but smaller, looser in general build, and lighter in colour.

No doubt the modern refined Hampshire-Down, derives its improved appearance from the cross with Jonas Webb's South-Down rams, effected by the well-known Mr. Humphries, of Oak-Ash, who hired three of the Babraham sheep at 60 guineas ($300.00) each for the season. Lamb-rams were first used by the same Mr. Humphries, for service, but with great caution, only 20 ewes being allowed to each lamb, though the number soon rose, as I myself have had nearly 60 ewes served by the same ram-lamb, and almost every one of them "stood". The modern Hampshire-Down is the heaviest of the Down breeds, and *au reste*, is only excelled by the Lincoln and though rarely, by the Cotswold.

My own draft ewes, four year-olds, used to weigh somewhere about from 96 to 104 lbs. the four-quarters, and that within a month or six weeks after weaning the lambs. Now, Prof. Wrightson will show any one who will visit the College at Downton, near Salisbury, ram-lambs that will weigh a hundred pounds the carcase in the month of July! Lambed in the middle of January, or thereabouts, the ram-lambs are commonly sold, for service in August and September, i.e., at seven or eight months old.

The old-fashioned ugly head of this breed, with its offensively prominent nose, has almost entirely disappeared under the influence of the Sussex cross. The face can hardly be too dark in colour, but the wool must be white; a dark tinge round the poll is fatal, and the ears must be free from any marbling and rather long, with a tendency to fall outwardly, which gives an air of width to the poll. The head has a good cover of wool both between the ears and on the cheek. As to the fine wool and the Roman-nose, they are as clearly hereditary and derived from the old horned sheep, as colour and the quality of the flesh are derived from the South-Down.

"Knowing the susceptibility of breeders," says Professor Wrightson, "it may be well to state here that when we refer to the mixed characteristics of the Hampshire-Down, we do not mean to cast any slur upon the breed as it now exists. The Hampshire Down has been too long established, and too long bred *inter se* (within itself), to be now charged with being of mixed origin. Every race of sheep has been crossed, with the exception of the Southdown, and possibly the Leicester."

No sheep does better within the hurdles than the sheep we are now considering, that is, unlike the Cots wold, he will stand being folded on rape, turnips, &., in large numbers. I remember well the numbers of the Western flocks, as they left the fold, lambs and ewes, on the hilly Downs near Andover. In Kent, Surrey, and other S. E. counties of England, the lot use to go to fold at sunset, on the fallows, and never leave it till the dew was off in the morning; and with nothing to eat all that time, they had to pick up their living on the downs in the day-time as well as they could.

The stocking of the down-farms, though, is now very different. Many more sheep are kept on a hundred acres than were kept in my day, for the method of feeding has been entirely changed. Every autumn, I used to go to Ewell or Guildford fair, in Surrey, to buy seventy or eighty three-year-old Hampshire-Down wethers to fatten for the use of my father's household. And such mutton it was! Now, people have to be contented with *tegs*, i.e., 12 to 15 months old sheep, but owing to the way in which they have been fed from their early youth, their meat, I am told, is by no means to be despised. But more of this when we come to the feeding of the flock.

When speaking of the useful properties of the Hampshire-Downs, it must not be forgotten that among them is their value for crossing. The Oxfords, you will remember are the descendants of a Hampshire-Down ewe and a Cotswold ram, and all the prize-winners in the cross-bred classes of the Smithfield Club Show in London, have had a Hampshire-Down on one side or the other.

Chapter Three - Oxfordshire-Downs

THESE, in reality middle-wooled sheep, which, for brevity's sake we shall here call "Oxfords", are, as we said at the end of the last chapter, the product of the Hampshire-Down ewe by a Cotswold ram.

The rule in breeding, as we said in the introduction, is to put the best bred a-top, that is, that the ram should, if there is a distinction between him and the ewe, the higher bred of the two. In this case, no one can doubt that the blood of the Cotswold is more free from foreign alliance than the blood of the Hampshire-Down; but there is more in this arrangement than a question of blood. The Hampshire-Down ewe of the period when the cross was first attempted, 1836 or 1837, was about the best nurse of all the ewes then bred, and that alone must have been sufficient to induce Mr. Samuel Druce, of Eynsham, to select her as the nursing mother of his future flock.

Mr. Druce's own account of the production of the Oxford by the crossing of Cotswold and Hampshire-Down blood, is worthy of attention. For though he speaks of them as got by a Cotswold ram out of a Southdown ewe, it is pretty clear that at that date, 1833, the term Southdown included all the Down-breeds from Sussex, through Hampshire, into Wiltshire; in fact this was the case, to my own knowledge, as late as 1840. In his letter to Mr. Philip Pusy. afterwards President of the Royal Agricultural Society, and one of the most practical of all the "gentlemen-farmers" I ever met, it is true Mr. Druce speaks in this loose manner; but in a subsequent letter, addressed to Mr. W. C. Spooner, he is more precise, and speaks of the cross as follows: "The foundation of this sheep was begun about the year 1833, by using a neat, well-made Cotswold-ram with Hampshire-Down ewes." Even at that date, the Hampshire-Downs were nothing like what we see nowadays, for it was subsequent to this that Webb's rams were used by Mr. Humphrey. They were, probably, what I recollect them to have been when I first visited the Surrey fairs; loose built, Roman-nosed, big, upstanding sheep, with plenty of black about the head, and with no particular merits of form about them; but hardy good-do-ers, carrying a vast amount of lean-meat about them, particularly

Oxford Down Ram

on the saddle, and good "butchers' sheep", i.e., with plenty of internal fat and heavy pelts.

The Oxfords were first called by the mixed name, "Down-Cotswolds," but before long acquired their present name of Oxfords; for the *shire* has been pretty generally dropped, and so much the better. Not many years ago, the cross in the blood was still easily distinguished by the mottled nose; indeed, we have seen it in pens of the Oxfords at Mile-End, within the last ten years; but at the last show held there, the muzzles of the Oxfords exhibited were of a uniform brown shade.

In the early days of the Oxfords the carcase weight of these sheep was thought satisfactory if, at from 13 to 15 months old, it came up to 76 lbs. Prof. Wrightson says that, nowadays, an improved Hampshire-Down, from a good flock, should weigh from 85 to 95 pounds, carcase-weight, at ten months, and that without forcing; and that at Britford fair, on August 12th, wether lambs are often seen to fetch 60s. ($14.40) a head, he himself having sold 100 on that day, in 1883, for 73s. each ($17.52); but prices for meat, at that date, were higher than they are now, though at present prices a 96 lbs. lamb — more properly teg — would be worth 72s. What used to take 13 to 15 months to bring about is now done in 8 or 10, for the modern principle seems to be to feed the land by feeding the sheep, instead of using so much artificials, and the quantity of cake and pulse and grain given to the flock is something that would make the Culleys and the Bakewells stare.

Shropshire Shearling-ewes (Hillhurst flock.)

Chapter Four - Shropshires

WHEN first we made acquaintance with the Shropshires — Shropshire-Downs they were then called, which w-as a mistake, as there is no such thing as a *down* in the old county — it was at a farm close to the town of Shrewsbury, the capital of the Shire, belonging to a Mr. Beach, in those days a noted breeder of these sheep, which were then, 1852, just beginning to win a high position in the English country exhibitions.

The history of the Shropshires is rather complicated. There are two old breeds on which the present Shropshires were engrafted. Oddly enough, too, these old breeds are natives of the two extreme points of the West Midlands, Cannock Chase at the eastern, and Clun Forest at the western extremity.

Several decades ago, there were developed great industries in coal and iron at Wellington, Coalport, and other districts in Shropshire. Wolverhampton, simultaneously, largely increased in population. The demand for mutton and lamb, of course, largely increased at the same time. To meet this demand, and to take advantage of it for their own profit, the farmers of Shropshire extended their turnip-and green-crops, and looked further afield for breeding sheep. The native stocks, in short, were not equal to the demand.

Breeding sheep were sought and bought in the midland and southern counties every autumn for many years, and they were walked to Shropshire and Staffordshire by thousands. Numbers of farmers paid this annual southward visit with this view. The occupier of Patshull at that time, Colonel Jones, was a pioneer in this movement. Some farmers bought Leicester ewes, others

Southdown, and others Hants-down [1] ewes. while, according as taste ran for an increase of wool, or early maturity was required, so Long-woolled rams were put to Short-woolled ewes, or the opposite practice was pursued. Thus, Shropshire became filled in the course of time with a large stock of all the best breeds of sheep in England. So much was this the case, that ultimately there was no necessity for the farmers of the West Midlands to turn southwards in search of stock-sheep. There still remained flocks of the old native breeds. Eventually these native breeds and the migrated stocks were brought together. Hence, the want of uniformity in colour, quality, and length of wool that existed thirty or forty years ago. And hence, too, this breed of sheep, like the Anglo-Saxon race of mankind, is equal to every quality of food, and adapted to almost every climate.

The stock of the old Cannock Chase sheep has no doubt given this breed the fine dark colour and fine flavour of their flesh. We have been informed that the flock of Beaudesert is the oldest one of this breed which has a recorded history. The quality of their flesh and fat has been celebrated for many years as being more like venison than mutton. So much was this the case that the late Marquis of Anglesea had unlimited standing orders from the distinguished guests who visited him to send quarters, sides, or carcases to noblemen and gentlemen all over the kingdom, and could his agent have produced ten times as many, the demand would not have been supplied. These were somewhat leggy and flat-ribbed sheep, with black points, *and some of them had short horns curving prettily upwards.* They of course took some time to get fat, and the mutton the noble marquis used to put before his guests was four or five years old. But so much for quality.

Then, there were the shorter-legged and more early maturing stock which had been cultivated in and around Clun Forest. The sheep undoubtedly had — and the old-established flocks still have — a large strain of the Welsh breed. Their contour and walk still show this. The ewes of this breed are bought in large numbers for producing fat lambs near London on the Essex and Herts sides. They are reputed to be the most prolific in yielding milk of any known breed. So well is this reputation established in the districts named, that the farmers do not mind losing 5s. a head on the ewes when they are sold out fat in the following summer, as they produce such good and early lambs that they make from 35s. to 45s. and 50s. in April and May. Lambs sell best in the London market when "spring-saled" has come him. This is, no doubt, the reason why Shropshire ewes may be justly looked upon as equal to any breed for suckling their lambs.

It was among these two breeds that the Leicesters and Downs, as above described, were introduced. Of course great want of uniformity and type was the result. Different opinions and tastes on the part of farmers had also much to do with this. Some preferred the old-fashioned mottled face with a South down type, while others liked larger sheep and black points. All this want of uniformity was made more and more conspicuous when the Shropshire breeders prevailed on the Royal Agricultural Society's authorities to appoint separate classes for the Shropshire breed of sheep. Judges at shows of course also differed in opinions. One year, two out of the three were in favour of the more Southdown colour and type, while the next year, two were in favour of dark colour and more size, notwithstanding the legs of the sheep were a little longer, and that the latter required more cake and corn to mature them early, or more time to get them fat in the ordinary way. The advocates of the latter argued, that there were several breeds of small sheep, some of which were deficient of flesh as compared with the fat they produced. Upon this they said "we have in the Shropshires large frames and ample lean of a dark rich colour. The smaller Downlike frames must be discarded, and the larger sizes cultivated." The results, as seen at the present time, have clearly proved that the latter advocates were right.

This conflict of opinions and diversity of taste led to warm discussions. It was shown that in more than one instance pure Southdowns had been introduced to flocks of the established Shropshire stock. In each instance the flock "went all to pieces," as it was termed. This was a lesson for the possessors of flocks which has been cultivated for many years on the lines above described. Out of this discussion, too, came the conclusion that dark points of uniform colour, with the largest possible size of frame, were the correct objects to arrive at. The experienced and consistent breeders came to this conclusion among themselves about the time of the "Royal" Battersea show in 1862, and

Shropshires Rams

most admirably have they carried it out by their skill in the art of selection.

It may seem odd at first sight to some breeders to read of a uniformity of black or dark faces and legs, when it is allowed or asserted that strains of the white-faced Leicesters have been introduced into the flocks; but this is just the point which throws a light on two leading features connected with breeding on the skill of the modern flockmaster, and on the way animals of a mixed breed will "breed back" from the strains of their ancestors of many generations ago. Take the latter point first: it occasionally happens in the best flocks of Shropshires that a lamb appears with a long, wavy, "open" or "watery" fleece. This is a clear indication that Leicester or some other Long-wooled breed was introduced to the Shropshire flocks at some remote period. The symptom appears as scrofula or other blood poisoning does in the third or fourth generation of mankind. The way, however, these "open" coats have been made exceptional brings us back to our first point — viz, the skill of modern flockmasters. When the lamb appears to be a half-bred, with a mottled face, whether it be male or female, it is at once discarded from the flock and fed for the butcher. In this has consisted the judgment, care, and skill of the modern breeders of Shropshires, who have brought their flocks to their present state of uniformity.

There are six or seven leading breeders whose names may be mentioned, as they have been so consistent among themselves that their flocks are nearly all alike in uniformity of type and general character. These are Messrs. Crane & Tanner, Shrawardine; Messrs. Minton, Montford; Mr. John Evans, Uffingon (all of whom live near Shrewsbury). Then, there are Mr. Thomas Mansell, Haringson, near Shifnal, and his son at Dunmaston, near Bridgenorth; Mr. W. J. Hock, Sutton Maddock; and Mr. T. Feen, Downton, who believes in size. Mr. John Darling, Beaudesert, near Rugeley, is now the possessor of the descendants of the Marquis of Anglesea's old Cannock Chase flock above mentioned and he is showing much spirit in endeavoring to develop it so that it shall be second to none, either in Staffordshire or Shropshire. Mr. Joseph Beach, too. The Hatton, Breewood, near Wolverhampton, inherited a flock that has been bred on the lines settled down upon by the older breeders above mentioned. I remember having a conversation with the late Mr. Joseph Beach some fifteen or sixteen years ago, when he was enthusiastic in favour of the larger size and uniformity of colour. The way this flock has been proved by selections is alike creditable to father and son.

It was indeed long before the Shropshires were admitted to the honour of a separate class. I remember well their first appearance in the ring; it was at the Gloucester show of the Royal, in 1853, but they were then lumped in with other sheep, as "short-wools, not being Southdowns." According to Prof. Wrightson, it was not till i860 that they were assigned separate classes, the type being then supposed to be fixed, and it was not till they had enjoyed the honour for two years, or so, that the Oxfords and the Hampshire-Downs won the same privilege. As a rule, the Shropshires head the list of the sheep classes at the Royal shows, as far as numbers go; at the Windsor exhibition of the Royal there were of

Leicesters	41	entries
Border-Leicesters	31	"
Cotswolds	60	"
Lincolns	58	"
Oxfords	82	"
Shropshires	212	"

32

Southdowns	123	"
Hampshire -Downs	78	"

In the Royal Agricultural Society's magazine for 1853, in the notice of the show of that year, we find the Shropshires thus described:

"They have no horns, faces and legs are gray or spotted; neck thick, with an excellent scrag; head well shaped, neither small nor large; breast broad and deep; back straight, with good carcase; hindquarters not so wide as the Southdows; leg straight, with good bone. They are very hardy, thrive well on moderate keep, and are readily prepared for market, the tegs weighing on an average from 80 to 100 pounds each, the carcase. Thus the Shropshire sheep, as contrasted with its maternal ancestor which grazed on the Longmynd Hills, had during sixty years doubled its dead weight." Writing in 1858, Prof. Tanner says: "Only a few years since, any mention of the Shropshire-Down sheep raised an enquiry, even among intellectual farmers, as to their character, few knowing anything about them." How altered is that state of things to-day! Every one interested in farming, from the plains of Australia to the sea washed rocks of Galway, knows the Shropshire; he is valued everywhere as a thoroughly trust worthy sheep that will do on poor keep and amply repay his owner for any extra food bestowed upon him; in fact, if I were not Alexander I would be Diogenes; that is, if I were farming, and could not get Hampshire-Downs to breed from, I would take up with Shropshires.

Suffolk Downs

Any one travelling through the Eastern counties of England, from London to Norwich, in the early thirties must have observed, unless the deadly dulness of the country sent him to sleep, on the borders between Suffolk and Norfolk, a number of dark-coloured, rough, long-legged sheep feeding about on the barren heaths that border on the two counties. They were the first "wild-sheep", so to speak, I had ever seen, so it is no wonder that, accustomed as I was to the smoother, more comfortable-looking flocks of the Southern counties, the Suffolk Health-sheep should have fixed themselves ineradicably in my mind. How great was my astonishment,

Suffolk Ram

then, to see by the reports of the Royal's meeting, a few years ago, that these rough, aboriginal sheep, had been so far improved that they had not only been taken up by some of the leading breeders of East-Anglia, but that, owing I believe to the exertions of the late Marquis of Bristol, the Suffolk-Down had been admitted to a place in the Royal's annual exhibition.

Arthur Young, in his tour, mentions these sheep, and from his account it would seem that, as long ago as 1790, they were considered to afford excellent mutton, though the wool can never have been good for much. However, as far as I can learn, not having the flock book at hand, there are six volumes of it extant; the Southdown has been the great medium of the improvement of the Suffolk; but whither have the horns vanished? for horns they certainly had when I saw them 65 years ago.

The Suffolk Sheep-Society thus describes the Suffolk; according to the points laid down by the best judges of the district:

"Head hornless; face black and long with a reasonably fine muzzle, especially in the ewe; ears a medium length, black and fine in texture; eyes bright and full; neck moderately long and well set on; shoulders broad and sloping; chest wide and deep; back and loin long, level, and well-united; tail broad and well set up; ribs long and springing well from the spine; legs and feet straight and black, with fine flat bone; woolled down to the knees and hocks; fleece moderately short."

A Dorset Horn flock (Hillhurst farm.)

The face of the Suffolk is as black as the Scotch sheep known as the "Black-faced;" it has no woo on the head or between the ears. In my younger days, there was a butcher, named Allen, who lived at the "corner of Mount Street and South-Audley Street, close to Hyde-Park, London, who killed no other sheep than the Suffolk, and made his fortune by them. He had a very traitorous animal in the form of a wether, who used to go with his drover to bring the sheep home from old Smithfield market. Now, all animals have a dislike to enter a slaughter-house on account of the smell of blood; in order to induce the market-sheep to enter, the wretch used to walk in at the head of the drove, and of course, as sheep always follow their leader, the rest, in perfect innocence, entered the slaughterhouse, and when they were all safe inside, the half door was closed; immediately on hearing the bang of the door, the villain bounded on to the backs of his betrayed friends, jumped over the half-door, and, leaving the other sheep to the mercy of the knife, gravely looked up in the drover's face in expectation of his reward which, in the form of a piece of bread, was awarded him at once.

A saddle of one of these sheep was worth eating.

The Dorset-Horn

No one who has ever seen a well-bred Dorset-Horn can mistake a specimen of the breed for any other kind. All other Down sheep have short wool and black or brown faces and legs, but in the Dorset we see a survival of a white-faced, horned, short-woolled race that may have inhabited the chalk Downs of Dorsetshire for many a hundred years before Caesar landed on the shores of Kent. No doubt the old Wiltshire horned sheep and the Dorset were nearly related; but the Wiltshire was early crossed with the Southdown until most of its characteristics disappeared, whereas the Dorset, as far as we know, has had no cross at all, but has been brought to its present state of perfection by selection alone.

Why Dorsets should differ from all other breeds of sheep in the time of their bringing forth their lambs, is a puzzle; all we can say is that the first breeders of the present stock probably put the ewes to the ram earlier and earlier every year, until it became a habit with them to "seek the male" betimes. For a Dorset ewe will take the ram in April, if she gets a chance, so that parturition will take place in September, and allowing, as we used to allow in England, the lamb to be killable at three months, Christmas and New Year's tide will not have passed before it will be on the market, and sell, at least it used to sell in our time, for a guinea ($5.00) a quarter.

The ewes, from what we gathered from a butcher who used to kill our sheep on the "home-farm," —as the farm is called in England that is kept in the landlord's hands for the supply of the house — the Dorset ewes, we say, used to weigh about twelve stones, or 96 lbs., when killed after having borne the last crop of lambs, i.e., at about five years old.

In 1840, Youatt described the Dorsets as entirely white, the face long and broad, the shoulders low but wide, back straight, chest deep, loins broad, legs longish, and the bone small. They were esteemed good folders, yielding well flavoured mutton, and averaging at three years old, from 16 to 20 lbs. a quarter.

What a change has taken place during the last 50 years! In 1890, Mr. John Kidner's first prize wethers, at the Smithfield Club Show, weighed, live weight 224lbs. each, say, 36lbs. a quarter; the wool, in 1840, weighed, on the average, 3¾lbs. a fleece, whereas, now, the fleeces of the ewe flock will run from 5 to 6 pounds, and of the rams from 8 to 12lbs, these weights being taken after the fleece has been washed on the sheep's back.

"Dorset ewes are capital mothers, and more prolific than any other breed of sheep. They will take the ram at almost any season, and if well fed before the time of copulation, will often bring two and not very rarely three at a birth. Very few of the rams survive their lamb-hood, but, according to the invariable practice in England, they are almost all castrated ten days or so after birth; but we must make one exception to the general rule: the Hampshire men used, in our time, to leave the male lambs as nature built them till the month of August, when the operation was performed, and happy is the man who visited the county in those days, for he stood a chance, at any of the country inns, of tasting that delicious dish emphatically designated "lamb's fry" not the liver, &c., but the "orchids." We dined off it, we remember at Wallingford, the day Andover won the Derby: in 1854, if our memory serves us.

There used to be plenty of Dorsets in the Isle of Wight, whence come the early supplies of lamb for the London market; Ireland has a good many, and, of late, several consignments have been received in Ontario, Mr. William Rolph being the chief importer. In 1889, Mr. T. S. Cooper, of Pennsylvania, landed 153 Dorsets, including all the first-prize winners of the Royal show of that year.

The earliest, or house-lambs as they are called, are treated in a peculiar way. A barn is set apart for the purpose of rearing them, and no expense is Spared. The building is divided into "Coops", in which the lambs are kept in separate lots according to their ages. Every evening, the ewes are turned into their young ones who speedily recognise each its own dam. After passing the night together, the ewes are sent to a fold of rape or turnips in the morning, *after the dew is off*. About ten o'clock, the ewes whose lambs have been sold, are driven into the barn, and held while the other ewes lambs empty their udders. At noon, the real mothers are driven into the lambs again for an hour or two, and at four the poor foster-mothers are again drained. We were told, many years ago, by a skilful practitioner of this plan, that the grand secret of success is to keep the barn at a regular temperature, any material variation of temperature, particularly upwards, being always attended by a serious loss of life among the lambs.

Of course, there are a multitude of other breeds of sheep besides those we, wath the assistance of professor Wrightson, have tried to describe, but as they are chiefly local, and by no means likely to be met with in this province, we do not think it necessary to enlarge upon their characteristics.

[1] Short or Hampshire; but the proper name of that shire is; the County of Southampton.

Chapter Five - Management of the Flock

IT may be laid down, as a general rule, that in the Province of Quebec every farmer, that keeps sheep at all, breeds his own. In England, we may say *en passant*, it is not the case; many farmers, especially in the hilly district of the north, breed from large flocks of ewes, but as they never keep male lambs after the month of September, it is clear that other farmers, those who buy their lambs, do not breed but feed. And the same is the practice in the Down counties; in Sussex, Hampshire, Wiltshire, &c., hundreds of thousands of well-bred wether lambs are to be met with at the Autumn fairs, the breeders only retaining the ewe lambs to refresh the flocks when the older ewes are drafted, generally at the age of four or five years. These the farmers of the turnip lands buy and fatten, after taking one crop of lambs from them.

But, here, it is not so. In the great majority of instances, what does one find to be the custom as regards sheep-breeding? We regret to say that if there is any animal on the farm that may be said to be neglected, it is that valuable animal, the sheep.

Originally, yes, up to very late times, sheep were kept in this province for the sake of their fleece; as for the meat they afforded, that was quite a thing that did not matter; it was used somehow or other in the farm-house, or perhaps a few lambs were sold to the butcher; but what lambs? Puny things that were never fattened, that, at best, weighed twenty pounds or so the carcase; how should they weigh more, when the ewe herself had to live on what she could pick up in the corners of the field, in the bush, and on that pasture, or *pacage*, to be found on the stubbles after three years successive grain-crops? Did the ewes die of old age, or what became of them? At all events, if they proved barren and were killed, who could have eaten their flesh; as for fat, it was non-extant! Still it goes on; every fall we see, under our windows in Sherbrooke street, troops of lambs, on their road to the butcher's pastures; ram-lambs, almost invariably, with their tails uncut and their *orchids* where nature placed them!

Again; what is the meaning of this tendency to put ewe-lambs, perhaps not more than six or seven months old, to the ram? We remember well when we expressed surprise to Mr. Mark Dawes, of Lower Lachine, at his following up this practice, that his reply was: "Oh! it would not pay to keep them a whole year doing nothing." If ample frames are to be perpetuated, females of all breeds or races must be allowed time to mature those frames, before being obliged to submit to the pains of travail, after having had to support another individual for the five months of gestation.

Consequently, it is not consonant with the laws of breeding to put ewes to the ram before they have obtained the age of eighteen months, so that they may bring forth for the first time when about two years old. Not very many years ago, it was the practice in England to postpone the first parturition of the ewe till she attained the age of three years; but that was in the days when ewes were allowed to get along as they could; now that ewes are carefully treated and well fed, the universal practice in all that country is to put the ewe to the ram at the age we have indicated.

And, now, let us consider how we are to celebrate the nuptials of the ewe and her mate: any how? By no means. Several questions have to be asked, and answered, before we can decide.

First, do we want our flock to produce lambs that shall be an improvement of themselves in frame or fleece, or in both? If so, it behoves us to be very careful in the selection of the ram. If the fleece of the ewe has a tendency to be loose and open, select a ram with a close wool. Should the ewe fall off too rapidly from the rump, choose a ram whose rump continues level as far as possible towards the setting-on of the tail. In fact, not to delay our readers too long over this matter, select a ram that will correct the faults perceptible in your ewes. As for the age of the ram, that will depend entirely on circumstances. If bred by a man who knows his business, and well-fed from its earliest days, there is no earthly reason why a ram-lamb of from seven to eight months old should not perform the functions of its office for, at any rate, 40 ewes. The practice is common in all the improved Hampshire-Down flocks of to-day as it was in similar flocks in 1850, and if the continuous practice of 50 years among the best breeders of sheep in England is not enough to warrant its soundness, we cannot see what is enough.

Do you want your ewes to drop twins? Well, it is not to be done by keeping them on short rations; they must be prepared for the service of the ram by careful feeding for at least three weeks or a month beforehand, and the best food for that purpose is rape.

How many ewes should be assigned to one ram? That, again, depends upon the age of the ram. A well-fed shearling ram can easily serve 60 ewes. Now, it is a matter of great importance that the time of service should extend over as short a period of time as possible; and, for this reason, the shorter will be the time from the first lamb dropped to the last; and if every sheep-breeder had attended the nightly lambing fold for five weeks, as we did in 1853, when, for what reason we know not, our ewes occupied all that time in getting through their work, they would spare no pains in shortening their weary labour.

There is one method by which the time of service may be prevented from extending over too long a period: use a teaser. A teaser is either an old worn-out ram, or a ram-lamb — the latter for choice — that is turned among the ewes but is prevented from service by a piece of sacking sewn on to the wool of his shoulders and extending so far down between his legs that his amatory propensities are baulked. Cruel, of course, but effective. The poor thing wanders about among the ewes, exciting their passions, and that so effectively that, when the Sultan himself is introduced into their society, the amiable creatures submit to his embraces without reluctance, and many services in the first four-and-twenty hours are the result, 25 not being an uncommon number.

As it is highly desirable that the ram be not exhausted by his arduous, we were always accustomed to put the ram with the ewes at about 4 o'clock in the afternoon; he remained with them all night, and at 8 o'clock in the morning he was removed to a small hurdled off space, away from his spouses, where he was well fed on cake, grain, and peas, with a little green-meat — clover, vetches. &c., — and restored to his sultanas at 4 p.m. as before.

In order to know what ewes have been served, the ram should be smeared, on the brisket, night and morning, with "ruddle", scientifically called "sesqui-oxide of iron", and all the ewes that have the red on their backs should be drafted into another field. In about three weeks, those that have not stood to the service will return, as it is called, when the ram should be re-introduced to them, and that will end his labours for the season.

And how shall we divide our flock of ewes? In accordance with their age, of course. Those that have already given birth to lambs will, or ought to go in one lot, those that are pregnant for the first time in another, and for this reason: the elder females must of necessity be fairly well kept during the winter, but the first year's ewes must be kept still better. Not, by any means, that they should, either of them, be made fat, for besides the fact that fat ewes almost invariably produce small lambs, we must not forget that the parturition of fat females of every description of stock, as well as of human beings, is always attended with danger. Keep the ewes, then, in good condition, but not too full fed. A little clover-hay, some peas-haulm, a few oats, and water at command, will see them safely through the winter.

In England, at least in the parts of England with which we are best acquainted, it is not the custom of shepherds and others connected with the breeding of sheep to talk of a two-year-old or three-year-old, but the age of a sheep is mentioned as connected with its teeth, as thus:

A teg, in some parts a hog or hoggett, is a yearling sheep before it is shorn. A two-tooth is a sheep from 12 to 15 months old until it, at 22 to 24 months, puts up another pair, when it becomes a four-tooth sheep.

At three years, it becomes a six-tooth, and at four a full-mouthed, with all its eight teeth complete. However, though this is a tolerably accurate account of the dentition of the sheep, it varies considerably from various causes. When a sheep begins to lose its teeth we used to call it a "broken-mouth"; in Cambridgeshire, a ewe that has lost teeth goes by the name "a crone", but in all well managed flocks it has long ceased to be the habit to keep ewes after the third crop of lambs; they are either fattened off by their proprietor, or else sold to the farmers of the non-breeding districts at the autumn fairs.

We should have mentioned that the reason why some farmers of apparently equally good soils do not breed sheep, but prefer buying from others that do not, to all appearances, occupy preferable farms, admits of a very clear definition: a farm, even all soils on any one farm, may not be suitable for sheep-breeding, but will fatten sheep very profitably. A very costly experience taught us this lesson, and we have never forgotten it: many years ago, in England, we took a farm of between three and four hundred acres, and invested a large sum of money in the purchase of ewes to stock it with; for it lay in a lovely exposure towards the south, and looked to be the very spot for breeding early lambs, more particularly as it had a chalk substratum under the whole it. The ewes lambed in the spring — lots of lambs, — but an overwhelming proportion of them were dead in their mother's womb; not only that, but eighty-five of the ewes died too; that is thirty per cent of the flock! And the only consolation we got was a question from one of our amiable neighbours: "Did no one tell you that on that farm the ewes always lost their lambs?" No one told us anything about it, or we should have done what we did afterwards, i.e., fattened off our ewes that were left and never bred another lamb on that farm.

A few turnips, or mangels, will do the ewes no harm, if the quantity is limited to five or six pounds a head daily, but beware of giving too many swedes, — a nasty experience of ours again — and mangel-*leaves* are said to be dangerous for in-lamb ewes, though why they are so we do not see. The great thing is to get the ewes to eat food that contains plenty of nitrogen J such as clover-hay, peas-haulm, a few peas as well, and a trifle of linseed-cake, commonly called *oil-meal,* which will tend to keep the bowels open, and produce a comfortable, satisfied feeling all over the ewe's body. Of course, we should prefer giving the seed of flax, crushed and mixed with chaffed oat or other straw; but that will come with time; at present, farmers have no linseed bruisers, and if the seed is ground between the stones, too much of the oil will be wasted, unless the seed is mixed with a large proportion of grain of other kinds, and that is not only troublesome, but renders it necessary to give a greater quantity of expensive food than need be. [1]

Sheep, just off the grass, do not care to eat dry food, such as chaff whether of hay or of straw; but as soon as the frosty mornings arrive, they will run to their troughs to see what they contain, and their breakfast will not detain them long.

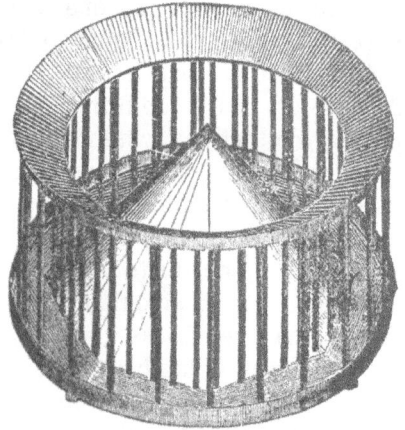

We cannot bear to see the little pains taken here, as a general rule, to preserve that valuable food, peas-traw or haulm. If the crop is not allowed to stand too long before being cut, and is carried without rain, peas-straw is, in our opinion, more valuable than timothy-hay, at least for sheep. But it must not be forgotten

A Model Sheep Rack

that, like hay made of vetches, one shower of rain on pease-haulm comparatively speaking spoils it, especially if it is not thoroughly dry before it is put into the barn.

It is astonishing how much cold sheep will bear provided their coats are free from wet. Good sheds will of course be provided for the ewes, and they need not be absolutely closed, as sheep will prefer lying about in the yard attached to the sheds even in very cold weather. Troughs, and roomy ones, for grain and chaff, should be numerous, but any common wood will serve for their manufacture. Pity that peas-haulm cannot be chaffed, but the quantity of sand and small grit, that is usually carried into the barn with it, blunts the knives of the chaff-cutter and renders the job not only tedious but expensive. A very useful rack invented by the well-known M. Eugene Casgrain is represented in the engraving: the following is a description of it, from the original in the *Journal d'Agriculture*:

"This rack is circular, and is made of two ranks of bars, with a hollow receptacle in the middle, in which is placed a cone, which makes the fodder thrown into it spread all around equally.

"The rack is 5½ feet in diameter and 4¾ feet high. There are twenty-two bars in the exterior rank, so that the same number of sheep can feed at the rack at once. The bars are 1½ inch in diameter, round, and so inserted in the sockets that they can easily be turned round and round. There is an interval of 7 inches between the exterior and the interior bars; the latter are twenty-three in number, one inch through, and 4 inches apart. Within the interior rank of bars is a wooden cone, 3 feet high, and 3¾ feet in diameter at the base. This cone, with the tackle that fastens the bars to the top of the rack, forms the receptacle for the fodder. A sort of raised shelf, three inches wide, is fastened to the top and bottom of the rack, around the outer rank of bars, and completes the whole.

"The advantages of this rack are these: its circular shape gives each sheep its own place without annoyance to his neighbours. For it is evident that they only approach one another as regards their heads, and the further one looks towards their hind-quarters, the more distant does each sheep get from his neighbours. This arrangement secure ewe-lambs and the in-lamb ewes from being hurt by being jostled by others. The bars of the outer range moving easily prevent the sheep from rubbing the wool off their necks. If the ewes pass alongside of the bars in a hurry, their mobility in the sockets prevents the eyes from being hurt. If the precaution is taken of placing the rack under a trap-door in the floor of the hay-loft, no rubbish can fall on the fleece of the sheep, which is thus kept perfectly clean. If, instead of fodder, roots are given to the sheep, the bottom of the rack, with its 3-inch-wide shelf, is there ready to receive the roots. The interior bars are near enough together to prevent the sheep from pulling out the hay or other fodder and trampling it under foot; and the distance between the bars, four inches, will not allow them to drag it out except mouthful by mouthful.

"An extra prize was awarded to this rack at the exhibition held at Montreal in the year 1883, and we think it *well* worthy of being reproduced here."

Notes. Don't feed your ewes, in winter, well one day and give them half rations the next: they will very likely gorge themselves the third day and perhaps abort: we have heard of such things happening.

We see in some of our exchanges that it has become a not uncommon thing to shear the lambs in the late fall: a dangerous practice with the ewe-tegs, for it is reported to us that many have died from the deprivation of their warm coat.

A dog in the ewes' premises should be shot at once; he has clearly no business there. He may not have come to kill, but his presence is enough to frighten the ewes out of their small wits, and but too likely lamb-slipping follows, though the ewe may go nearly up to her time.

For goodness sake don't let your in-lamb ewes wander about loose in the deep snow in February and March, at any time indeed. How often do we see ewes heavy in lamb in the road near the farmhouse; a sleigh passes, the ewes take fright, dart to the side of the road, plunge into the untrodden snow in the ditch, and, unless the man in the sleigh is merciful and helps the poor thing, her funeral peal will soon be ringing.

By the bye, we forgot to say, but every sheep breeder knows it already, that a ewe goes 5 calendar months with lamb, say 22 weeks, on the average.

For goodness sake don't grudge sheep in regard to litter: a dry bed is everything for a sheep of any kind, how much more then is it needed by the in-lamb ewe!

When do you intend to have your ewes lamb-down? That you must arrange according to circumstances, remembering one thing: ewes that have been accustomed, for their first two pregnancies, to lamb in April, will not, because you put the ram with them in August, oblige you by taking him at once; the odds are, that if you do so, the ewes will come into season one by one, the general impregnation lasting very likely five weeks, to say nothing of the bother of those who "return." You will not neither will your shepherd — if you keep one — find five or six weeks attendance, o'nights, in the lambing shed by any means pleasant. We always attended to our ewes ourselves, though we had a first-rate Sussex shepherd, and the climate in which we then farmed was by no means so stern as that of the province of Quebec.

A very sensible piece of advice is given by Mr. Casgrain, referring to the danger incurred by the ewes in jostling one the other in going through the door of the lambing-shed. He recommends that the sill should be raised fifteen inches above the floor, and that there should be no approach to it except by a little bridge, without rails, only wide enough for two sheep to pass at once; the exit of the ewes will take up more time if thus arranged, but it will not be so likely to cause accidents.

Prof. Wrightson says, with perfect reason, that "with half a pound of hay and ten or twelve pounds [2] of turnips, ewes during pregnancy require no cake or other expensive food. True; but that would not suit ewes in this country and climate; for the ewes he speaks of lambed in early January, and in fortunate Southern England there is a good bite for them in the meadows and on the Downs, all the time of their pregnancy. Here, we have no such advantage, the trifle of cake a day we recommend, 4 oz. a head, will not cost much, and the outlay will be well repaid by the good state of health it will help to ensure to the ewes.

Mr. Casgrain does not approve of the litter and dung, &c., being allowed to accumulate for months under the sheep, as, he says, is the practice of many farmers, and therein we agree with him, though in the severer months of winter no deleterious emanations can possibly escape from the bedding. [3] But he goes on to say that "the fresh manure will be better;" how it can be better that the other when the other has lost none of its good qualities, we do not see.

But we must not forget the case of the yearling ewes, just now in lamb for the first time. These will of course require better keep than the older ones, as will also the tegs, or lambs of the year.

[1] The fact that, out of 100 grains of flaxseed, given to cattle or sheep uncrushed, at least 80 will be passed undigested in the faces, should deter farmers from wasting such valuable food.
[2] But, as we observed before, beware of too many turnips.
[3] The pressure of the weight of the sheep will prevent any heating.

Chapter Six - Drafting the Flock

THE first operation after the lambs are weaned and the ewes are preparing for the introduction to married life, is the drafting of the flock. This, if you want your flock to gain any place at exhibitions, sales, &c., must be carried out in the most hard-hearted manner. But of, say, 60 ewes there will probably be

> 16 two-tooths.
> 16 four-tooths.
> 16 six-tooths.
> 14 full-mouthed, and older
> ___
> 60

and you will have no difficulty in finding among them at least a dozen which, for one reason or another, are not worth retaining in the flock. Of the older ewes, that is, of those who are over four-years old, some will be what is technically called it "broken-mouthed" or "crones", i.e., those that have lost one or more teeth; others will have shown signs of losing their wool; here, will be one that had not sufficient milk to supply the wants of its last lamb, &c., &c. Of the younger ewes, there will be some whose form is not what it should be; a weak neck in one, flat ribs in another, a weak loin in a third, Sec, will cause them to be discarded.

Then comes the selection of the young ewes to supply the place of the discarded in the following season; and after all this work of selection is over and done with, we have to go to work in the most important of all, the choice of

The Ram

Says Prof. Wrightson, speaking of the father of the flock: "A good and improving flock of ewes, mated with rams likely to still further correct their weak points and produce ewe-lambs which will again add to the reputation of the flock, is of the greatest importance."

"Rams must be bought with the strictest eye to personal merit. Success does not depend upon price, but upon securing good sheep. Go to the sales of the best flocks, but back your own opinion. A breeder will always find it expedient to be a buyer but he should always retain one or two of his own ram-lambs for service in his own flock, remembering, that it is not advisable to put them to ewes too near in consanguinity." Did we mention that Jonas Webb once told us he never mated ram and ewe nearer than sixth-cousins.

Notes: Mate strong ewes with rams of a more refined character, and the reverse.

A light-coloured ram should be mated with dark ewes.

If a preponderant fault of your ewes is want of width, or lean neck, or scanty legs, put them to a stout ram, with strong scrag and well-rounded "legs of mutton."

Rape for a fortnight or three weeks before coupling will bring the ewes into humour for the ram.

Prof. Wrightson's plan is this: after the ewes have all been served, he puts one ram with the whole lot of them so that those ewes that "return may be served again. That the number of twins depends a great deal on management, is clear from the fact that some breeders invariably have a much heavier crop of lambs than others. Of course, as the *rape* treatment is the almost invariable practice of the great breeders of Essex, Cambridgeshire, &c., and was always resorted to by those leading flockmasters, the four brothers Webb, Sam. Jonas of Ickleton, John Clayden of Littlebury, and the like, it is clearly not a fantasy.

Some farmers in this province have told me that they would rather have their ewes drop singles than twins. Well, if they grudge a little extra food, and the trouble of keeping these ewes and their twins apart from the ewes with singles, I have nothing to say about it; but I should be glad to know how they treat a ewe that has lost her solitary lamb? Milk her and make the milk into cheese, we suppose! If a ewe is properly fed, she can nurse two lambs without drawing too much on her fund of life. We never saw many ewes in the districts round Joliette, or in those round St. Hyacinthe, but the small flocks in those parts always seemed to us to be kept for no definite purpose, but rather because it was the custom to keep sheep and the owners did, as their neighbours were in the habit of doing. It is full time, now that mutton is of such great value in our foreign market, the last report from London quoting "Shorthorns, 4s. 6d. per 8lbs — South-downs, 6s." or, in our currency, and by the pound, 13$\frac{1}{3}$ cents and 18 cents — that our people should look about them.

Dipping

Our general practice here is for our ewes to lamb-down so late in the spring that there is not much danger of the foetus being injured in the operation; for the early part of September is about the best time for dipping ewes, and then the services of the ram has not, in nine cases out of ten, been put into requisition. In my time. Bigg's composition was the favorite sheep-dip, but the late lamented Sir John Lawes brought out one quite as good towards the latter end of his useful life. Mr. Henry Gray, druggist, of St-Lawrence Main-Street, Montreal, has my notion of what a sheep-dip ought to be, and I can conscientiously recommend my readers to apply to him when in need of the materials.

The best time for dipping, in my opinion, is just after harvest — say, the end of August — and the lambs, all of whom should be weaned by that date, should be dipped at the same time.

Sheep are dipped for two reasons: first, to kill ticks; secondly, to clean and promote the healthiness of the skin. It is, I hear, a common practice in England nowadays to dip sheep twice a year. In my time, once was considered sufficient, and as I never tried the second dipping I cannot speak of its effects from experience; but the English flockmasters know; they are not given to waste time or money, and if they do it, their example in this, as well as in other things, is a pretty safe one to follow.

Notes or points. A sheep's head, here, is not worth much, except in large towns, where there are plenty of Scots, and people that, like ourselves, who have been a good deal in Scotland, have learnt to like a tup's "head" converted into soup with plenty of onions, celery, carrots, and pearl-barley. And yet that very member, if it is perfect, or even fairly good, will add many a dollar to the selling price of a ram.

When I went to live with the great Southdown breeder, Wm. Rigden, to study his flock, he was deeply interested in the improvement of the necks of his sheep. He succeeded in his design, in about three years, by selecting a ram every year from the Babraham lot, which shone especially in that member.

Width over the crops, great girth, and thickness through the heart are all important. These cannot be given in a picture; but study the Hampshire-Downs, the Southdowns, and the Shropshires, at the next Provincial Show, and you will see what they indicate; among other things they are almost infallible signs of a good digestion.

Chapter Seven - Lambing Time

THE time is at hand I hope, when we shall see, under the care of shepherds, flocks of sheep carefully looked after, and fed during the whole of the summer on green-crops grown expressly for their consumption. I cannot conceive of any other means of restoring the fertility of the worn-out farms so common in some parts of the province. Sheep, even when the flock is small, pay their owner, for if they did not, we should not see so many let out on shares. But kept, as they ought to be, on the land, [1] from May to December, they will not only yield the ordinary return of a lamb and a fleece, but they will at least double the production of the farm.

As I said before, two things are specially to be sought for: twins, and a general lambing in the shortest possible period.

You may think yourself very fortunate if you have secured a good shepherd. I had one and only one, but he was a phenomenon; he knew personally every ewe in the flock, its descent, and when it was due to lamb; he knew how to assist it when it needed help, but he always let it alone when assistance was unnecessary; he never wasted the food set apart for the ewes, knew how to care for a sick ewe, how to rear a lamb that has lost its mother, and how to induce a ewe, with a single lamb, that had lots of milk, to act as foster-mother to a lamb whose dam had not enough. One never heard, in those days, in my lambing-shed, the baaing of ewes and lambs that had lost one another; his watchful care never grew slack, and during the time he was in my service, he discharged his duty like a man.

The number of ewes a ram can serve depends upon circumstances. A Hampshire-Down lamb-ram, 8 months old, can manage without trouble from 30 to 40; a shearling of any breed from 70 to 80. A Southdown shearling I hired from Jonas Webb served 110 ewes, but they were young and vigorous, in full condition, and great pains were taken by the shepherd to keep the ram well fed and to give him plenty of rest in the early mornings. To the best of my recollection, all but 25 of them gave twins, and one especial night I well remember, in which 20 ewes presented me with 39 lambs!

The Hampshire men prefer lamb-rams; but their ewes lamb so early, about the middle of January, and are so well fed, that their lambs are as strong in the months of August and September as are the shearlings — 2 tooth's — (Ungrammatical; but such is the technical form) — of other breeds.

As I said before, the ram should be put into a small lot, with one ewe with him to keep him quiet, and be fed freely with any food he seems to prefer; if he is thus kept away from the rest of the ewes for a couple of hours twice a day, it will be all the better for his service, as when he is allowed to run day and night with the ewes, he is so attentive to his duties that he will not give himself time to eat.

At the end of a fortnight, most of the ewes will have been served, indeed, if properly treated, every one of them.

Ewes that are too fat, almost invariably give birth to small lambs; sometimes, they suffer from inflammation of the womb in lambing; therefore, while keeping your ewes in good condition during pregnancy, don't let them get too fat. On the other hand, ewes that are too poor cannot nourish the foetus well, die of weakness if the lambing is difficult, lose their wool, and starve their lambs; so, don't let your in-lamb ewes suffer from hunger.

As I remarked before, in sheep, as well as in all kinds of horned stock, a trifle of oil-cake or crushed linseed has a marvellous effect in easing the pangs of maternity; and as it is so, no one should grudge his ewes a quarter pound of cake a day for a fortnight before and a fortnight after lambing; the cost will be a mere bagatelle compared with the benefits conferred.

Patient darlings! how mildly and sweetly do they give themselves up to the tender care of the shepherd, when the labour is hard! It is worth any one's while, whether he is a breeder or not, to watch a flock of ewes in the lambing-shed. He will never again believe that "We have no duties whatever to discharge towards animals", unless he is utterly devoid of all sympathetic feelings.

Mind that your ewes are kept perfectly undisturbed. The sudden apparition of a stranger may so terrify them that irreparable damage may be caused in a minute. Ewes can stand a good deal of cold, as can the lambs, but beware of a wet fleece. Open lambing-sheds answer very well; in fact, we prefer them to close sheds; but there should be means provided to compel the ewes to keep under cover when it rains, for, in very truth, amiable as they are, sheep "have not got sense enough to come in when it rains."

You will soon learn to distinguish from the others the ewe that is at the point of labour. The parts beneath the tail begin to grow red and swollen; she seems ill at ease, she wanders hither and thither, tries to get away from her sisters. In fact, she gets what would be called among men "fidgety", taking a great interest in the lambs of the other ewes, even trying to appropriate them. Then, the bag of liquid protrudes itself from the womb, and, if the presentation is a natural one, the two fore-feet of the lamb, with its muzzle resting upon them. The ewe shifts her position from time to time, gets up on her feet, lies down again, making many an effort to rid herself of her burden. Now is the time when, if the ewe seems weak, the shepherd's services may be of use: drawing the legs of the lamb towards him as gently as possible, and freeing the forehead from the pressure of the vagina with his fingers, he pulls the legs softly, in a *downward* direction, being careful to time his pulls with the efforts of the ewe. He must never pull between the pains, for unexpected aid always surprises the ewe, and makes her neglect to aid herself.

When the lamb has been successfully brought to light and is laid before its dam, she, unless the labour has been very severe, recognises it at once, caresses it and lavishes a host of tender appeals to it that none but a mother can appreciate. In the case of twins, you must take care that the second is brought forth as soon as possible, though it rarely delays its appearance; but it sometimes happens that the ewes is so taken up with the elder of the two that she does not attend to the pains caused by the younger. We have sometimes seen the second come to light without the ewe seeming to know anything about it.

In the case of a false presentation, the shepherd should introduce his hand, previously greased — goose-grease is the best, as it retains its oiliness longer than any other — into the womb of the ewe, and extract the lamb as gently, though as quickly, as possible. The Leicester ewe is said to need assistance at parturition more than other sheep, but in the Downs we never observed many cases of mal-presentation; perhaps a leg doubled-back at the knee-joint, or some trifle like that, may occasionally be met with; this is easily detected, and easily remedied.

Sometimes, after a hard labour, the ewe seems inclined to refuse to let the lamb suck; the udder should then be examined, and if it seems inflamed, it should be bathed in a weak solution of nitrate of potash — saltpetre — or with hot water alone. If it is found that the udder is neither inflamed nor hard, the ewe should be tied up tightly to a post, and the hind-quarters held firmly till the lamb has drunk its fill. It will not be long before the difficulty is overcome and mother and child on good terms. Should a ewe lose her only one, give her one from a ewe that has twins. To get her to take to it, strip the skin from the dead lamb and tie it over the foster-lamb, and the ewe after this is presented to her will mother it; if the skin has been taken off the corpse before it has had time to cool, the dodge will be all the more successful. A single season in the lambing-shed will teach you more of the best means of encountering these unavoidable troubles than all our description.

If a lamb just born seems reluctant to rise, look at its muzzle, and if there is any watery mucus round the mouth and nostrils, clear it away; it might stifle the lamb.

After a hard labour, bathe the uterus of the ewe with a very weak solution of carbolic acid.

No careful shepherd, accidents apart, ought to be satisfied if he loses more than 5% of his ewes at lambing-time; but he need not be ashamed of his work if he confines his losses to 3%. There is no one single thing on a farm about which so many lies are told as the losses in the flock.

The really good trustworthy shepherd can tell at a glance if any one of his ewes is ailing, and will know just what remedies to apply to relieve her: he loses no time about it either.

If you have more lambs than your ewes can bring up properly, the surplus may be reared "on the bottle," with warm cow's milk. But motherless lambs are always a nuisance, baa-ing about all the time, running into the farm-house after every one they see, and not unfrequently bouncing over the garden-fence and doing all sorts of mischief. The best thing to do with them is to eat them as soon as they are fit for the knife.

As for the castration of the males, practice varies. One thing is positive, every male that is not to be kept for breeding, should undergo the operation sooner or later. The Sussex folk cut their lambs at about ten days from birth; the Hampshire men let theirs go uncut till they are four or five months old, but both sets of breeders cut the tails as soon as the lamb is strong enough to bear the operation. Whenever the lamb is cut, choose a mild day for the job. We regret to say that the early lambs — the later ones, too— that are sent to Montreal Market are never cut. It may seem useless to cut such very young lambs as these, but there is a certain redness that appears on the flesh of un-cut lambs that certainly spoils the look of the meat and cannot but have some evil influence on its flavour. A fine lot of our lambs sold for but poor prices in London on account of not having been cut; this was in our very early days, in 1848. Our English shepherds dock the tails of the lambs much shorter than is the rule here, and, in our opinion, they are right. A tail cut short gives the hind-quarters a squareness to the eye of the buyer, which is decidedly tak-ing; and as the real reason for docking sheep is to protect them from the hanging on of filth, which leads to the attacks of the fly, that lays eggs which turn into maggots, the shorter the tail is, in reason, the better. About the third joint is the proper place to cut, but if you cut between two joints, it does not make much difference, as the part left on will slough off, if you tie a string tightly round the tail just above it.

If you wait, like the Hampshire men, till your lambs are four or five months old before you castrate them, you will have a chance to taste that delicious dish, delicately called "lambs' fry". Clean and split the orchids, but don't wash them; dry them carefully with a napkin, dip them into yolk of egg, and roll them in bread-crumbs, that must be finely sifted and quite dry, putting plen-ty of dry sweet-herbs, parsley, cherville, thyme, savory and marjoram, and then fry them in plenty of olive-oil till they are a fine brown. The fat or oil for this must be very hot — fat won't boil at ordinary pressure, though the water in the fat does bubble up.

***

The ewes and lambs should of course be put on the best pasture your farm affords as soon as it is fit for their reception, i.e., as soon as there is a fair bite for them; but continue to give them some dry fodder, if you can get them to eat it. As soon as possible, begin the sowing of vetches, rape, and other green-fodder crops, not forgetting that your mainstay for the lambs must be varie-ty. In the Southdown country, where fall-sown crops of course succeed, the lambs and their dams taste at least four different kinds of food daily: winter oats, winter barley, rye, and the pasture on the Down. We have not that ad-vantage, but we must do the best we can, and take care that there is no *inter-regnum* between the different courses served up to our flocks; for I wish it to impress on your minds that, of all young animals, the most difficult to bring back to a thriving state, after being once allowed to fall off, is the young lamb.

And what green-fodder crops shall we grow for our flock? Tares or vetches, mixtures of different grains and pulse, and above all, *rape.*

Tares, or *Vetches.* — This is the first crop on our list, and well known to every farmer. It will grow well on all soils, but prefers a clay-loam. On sands, or gravels, it demands a fair dressing of manure, but on heavy land, in tolerable condition, it can do without. As tares are inclined to fall down when they are at their best, it is customary to sow 2 or 3 pecks of rye, or oats, per acre with them, but as rye soon becomes uneatable, and horses don't care much for green oats, half a bushel of wheat, at present prices, would be worth trying.

The quantity of seed required is 2½ bushels to an acre, when the land is in fair order, but 3 bushels would not be too much in rough ground. There are two sorts, the winter and the spring tares; the seed of the former is small, that of the latter much larger, but the quality of the forage of the winter tares is so much superior to that of the spring tares that, in the East of England, they are sown invariably to the utter exclusion of the other sort. A couple of bushels of plaster to the acre, on the young plant, will materially assist the yield. It is well to observe that nothing is gained by very luxuriant crops of tares, as they always fall down and waste themselves, unless cut at the critical time of coming into bloom.

A very productive mixture for forage is: 1½ bushel of tares, 1½ bushel of oats, ½ bushel of horse-beans and ½ bushel of wheat. Of course, the roller must follow the harrow at seed time, or else the unhappy man who mows the crop will lose temper, and the farmer's time, pretty frequently. Two sowings should be made, one 3 weeks after the other.

The time will soon arrive, when we shall no longer see the sheep lying under the fences, and depositing their invaluable manure on the grassy borders of the arable lands out of the plough's reach. To the system of folding off green crops all the summer, as well as root crops all the winter, the formerly poor lands of the East and South of England owe their present fertility. Here, the end of October must, as a general rule, see the flock in the yards; but it is my firm conviction that Canada never will produce the full amount of grain which it is capable of producing, until the sheep is made, what it is in England, the dung-carrier of the farm. On the sandy soils of Bedfordshire, as well as on the chalky clays of Kent, towards the beginning of July, the traveller sees, as he journeys along the roads, large fields of a rich, green plant, something like at all cabbage stalk, with leaves growing all the way up it, and from 3 to 3½ feet high. This is *rape*, or Coleseed, the Colza of the French. The latter, distinguished from the former by the roughness of the leaves (hispid), is supposed to be, and, perhaps, is the more fattening of the two; but they will both make sheep *ripe fat*, without any other food. Cows are fond of it, and it makes them give plenty of rich milk, but great care must be taken that it is not given to them with the dew on, or in rainy weather. Insects don't trouble it; at least, used not to; but of late, I hear that the fly (*haltica nemorum*) does trouble it. As it is sown broadcast it requires no hoeing; and no weed can struggle against it. It is grateful for manure, but on good soils, of a moderately heavy quality, it can do without it. A few, say, 10 bushels of bones, mixed with as many bushels of ashes, lixiviated if no others can be spared, will, on light soils, produce a crop so luxuriant in its lush abundance, that no one can fail to appreciate it. If it is desired that the ewes should bring forth twin lambs in the spring, a fortnight, or three week of feeding on rape, before the ram is put to them, will have the wished for effect. The plant grows so high, and is so thick on the ground, that the sheep don't trample it down, as they do clover; for which reason they may be allowed to feed on it at liberty; though of course, the more economical plan would be to divide a small piece off, every two days, or so.

I submit a sketch of the newest kind of *Hurdle;* it is usually made of iron, but I have used some of wood, which answered perfectly. It will be easily seen that a boy can move them backwards, or forwards, without difficulty. If sheep are given to jumping, it would be a good plan to leave one of the upright bars of each hurdle 18 inches higher than in the sketch; if a wire is then run along the tops of the bars, loosely will do, the sheep may try to jump at first; but, after a few attempts, the shock they receive on falling back from their spring will so astonish their weak nerves, that they will become disinclined to further adventures. It is in this way alone, that the Welsh Mountain sheep the wildest domestic animals in creation, can be kept within bounds. The fresh piece should *always* be given in the afternoon, say about 2 o'clock, when the sheep will have their bellies pretty full, and the evening dews have not begun to fall.

One advantage that this hurdle, when used for ewes and lambs, has, is that while the ewes are kept confined to one place until the fold is advanced, the lambs can go forward into the, as yet, untouched vetches or rape. Of course, those who mean to push the lambs, will have a trough or two outside the fold, within which a few pints of peas, or mixed peas and oats, will be placed twice a day. Not only will the lambs be improved by this extra food, but the land will profit by the additional richness of the manure dropped by sheep. In England, many of the best farmers are giving up the use of artificial manure — always excepting superphosphate for the root crop — and devoting the money formerly spent in its purchase to the acquisition of maize, cake, or linseed, for their sheep and cattle.

***

51

The land should be as carefully prepared for rape as for mangels, or swedes. A cross-ploughing, in the spring followed by the harrow, the roller and the cultivator, or grubber, if there is one handy, should leave the land in a fine tilth by the middle, or end of May. From 6lbs. to 8lbs. of seed per acre can be either harrowed in, with light harrows, or with a bush; or, if the soil be a little cloddy, I roller may be used; but so early in the season a should prefer to leave a harrowed surface to a rolled surface, as being less likely to suffer consolidation and, consequently, hardening on the top, from heavy rains followed by a hot sun; unless, as in the case of tares &c., where a scythe has to be subsequently used, and even then I would rather roll after the crop is up.

Not many diseases trouble lambs, as long as they are on the milk. Diarrhoea is sometimes brought on by a change from a poor pasture to a rich one, but this generally yields to a dose of half an oz. of Epsom salts, with a little ginger to soothe the bowels. Of the contrary complaint, constipation, which until after they are weaned rarely affects lambs, a slight aperitive will usually relieve them. Care must be taken, especially in bush farms, to keep the hinder-parts quite free from filth, lest the fly should attack them. The best plan is to shear off the long wool that grows between the thighs, and then to bathe the lambs in the preparation we already spoke of. The head may be protected by a cap, as in the following engraving, but if there is the least wound on the forehead, no cap should be used, as the fly might have dropped its eggs beforehand, and the maggots would then be hidden from the shepherd's eye, so that he would not be in a position to deal with them.

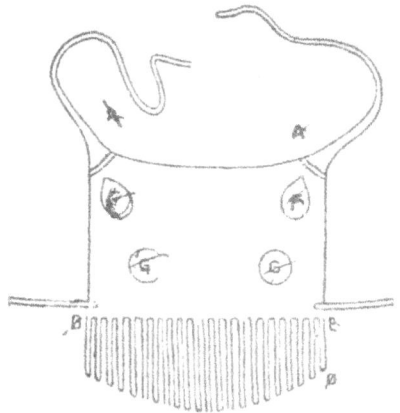

(1)

The cap (i) is made of lamb-skin, and is tied on the head as shown in (2). Stewart advises that it should be smeared with a resinated carbolic acid salve, which should be renewed every

(2)

week; but very few shepherds would take so much extra trouble.

[1] That is, no allowed to lie about in the bush or in the ditches, where their droppings will be wasted.

Chapter Eight - On the Weaning of Lambs

LAMBS are usually weaned at the age of from three to four months. Here, in the province of Quebec, they are generally left to wean themselves. In fact, the ewe often seems, at the above period, to get tired of nursing, and repels the lamb when he seeks food "from nature's founts." And it is not wonderful that she should weary of her work, when we see the terribly hard bunts her tender udder gets from the bossy frontal of her child.

It would seem at first sight but a simple thing to wean a lamb. One would say: Oh! take it away from the ewe, and the thing is done. True, but consider a little. Suppose, for instance the ewes and lambs are in a field, and you take the lambs away from their dams and put them into another field. What a jolly row there will be! The lambs perfectly strange to their new home, wander about, baa-ing after their mothers and lose flesh in their ardent search after their dams and their accustomed playground. It will take them a few days to get accustomed to their new life, Now, look on this picture: take the ewes away from the lambs, after they have passed a few days in one and the same field; take the ewes away to another field, out of sight and out of hearing of the lambs, and it will not be long before the young ones will be seen contentedly feeding away in perfect tranquillity. Those who have already weaned themselves, as we said just now, will be feeding, quite satisfied with the grass, and at that sight the feelings of the others will be soothed, and they will soon follow the examples set them.

If the lambing is early, the weaning should be early too; otherwise, the ewes will not have time to get into good condition before the season of coupling recurs. Only conceive that, even in our day in Scotland, the ewes were milked after the lambs were weaned, as I believe they still are in some part of France and in Piedmont! and all for what! for the gain in making a few pounds of cheese that some people appear to like! However, In Britain this is no longer done.

Though we object to the practice of milking ewes after weaning their lambs, we must watch over the ewes in regard to one thing. Every now and then, a quick-eyed shepherd will find that a ewe, whether from having lambed late, or for whatever cause, may have a flow of milk even after the weaning is over. In such a case, she must be dried off, exactly as a cow is dried off before calving. Milk the ewe at intervals of 12 hours, then of 24 hours, then of 36 hours, and we need hardly say that, for a week or ten days, the less succulent her food is the better. It is by neglecting these cases, in cows as well as in ewes, that so many of them lose one or more teats.

After an interval of separation of from ten days to a fortnight, the lambs and ewes can be restored to companionship; all filial instinct or maternal feeling will have utterly vanished.

Notes. Why should so many black sheep be kept in the French part of the province? Is it to save the trouble of dyeing the wool of the flock? Curiously enough, a very large proportion of the early lambs sent to Montreal are black. On the 27th of February I saw two very fine, though not large, fat lambs hanging up in Messrs Browns' shop, St. Catherine street. They were ripe-fat, the kidneys well covered, and the briskets full of meat, but they handled soft, and a few pints of peas, with a fortnight or so longer feeding, would have made them much better meat.

The following is the system of feeding the flock of one of the leading breeders of Hampshire-Downs in England:

Before lambing, the ewes have turnips and *sainfoin* hay; after lambing, they have a mixture of one part hay to six parts wheat-straw chaffed, with a bushel of pulped roots— swedes or mangels — to sixteen bushels of the chaff; to this is added half a pound of malt culms, and the same of peas-meal, to each ewe, the whole being thoroughly mixed and allowed to ferment for thirty-six hours. Ewes nursing twins have a pound of cotton-cake in addition to the above. *Malt culms* are the dried rootlets of the malt. Some times called *coombs.*

Mixed foods are clearly the best for all kinds of stock. Professor Wrightson, in a letter I saw some years ago, recommends the following as being composed of farinaceous and albuminoid constituents in fair proportion:

Advantage of Mixing Foods

"We certainly recommend a mixture of concentrated foods. Linseed cake alone is too heating, and if the sheep are to be kept in health it ought to be mixed with a proportion of foods poorer in albuminoids. If this precaution is neglected we shall run a chance of sore teats and sore mouths. In the above remarks we were chiefly aiming at arriving at the limits of cost. We now suggest that a mixture should be made on the most economical and scientific grounds possible. The mixture should be readily constructed, and be free from complication. It should be composed of farinaceous and albuminoid constituents in fair proportions. We suggest the adjacent: —

It is pity, sainfoin is not more

First mixture for ewes or tegs :—

1 bushel of	linseed cake...	}	1 to 1½ lb. per head.
1 "	cotton cake....	
2 "	maize (crushed)	

Second mixture for ewes :—

1 bushel of	bran..........	}	1 to 1¼ lb. per head.
1 "	linseed cake...	
1 "	barley........	
1 "	maize	

Third mixture for lambs :—

1 bushel of	white peas.....	}	Quantity p. head to vary with size.
1 "	linseed cake...	
1 "	malt culms....	
1 "	crushed barley.	

Fourth mixture for dry sheep :—

1 bushel of	wheat........	}	1 to 1½ lb. per head.
1 "	barley........	
1 "	oats..........	
1 "	linseed cake...	"

cultivated here. I grew some, on Mr. Dawes' farm at Lachine, some ten or twelve years ago; it did very well, but, unfortunately, the land was wanted the second year for some other purpose, and the sainfoin was ploughed up before it had an opportunity of showing what it could do at its best. I say, it is a pity it is not grown here, for there is nothing equal to it as a pasture for weaning lambs. It grows well on all calcareous soils, and lasts for several seasons without renewal. I have never seen lambs scour when on sainfoin, though I have often seen that malady affects them when on red-clover, and these latter patients were completely cured when removed to a sainfoin-lea.

As I mentioned *sainfoin* just now, I might as well say a few words on its cultivation, etc. The seed is generally sown in the capsule, if *milled* seed is sown, 40lbs. to the acre will be sufficient. We see, upon looking back into one of our journals that the sainfoin we grew at Lachine was fit to cut 17 days before red-clover. For hay, it should be cut just as the blossoms are beginning to expand, and as it is a very early crop, it would be well, to increase the first year's yield, to sow with it about 5lbs. of yellow trefoil, *medicago lupulina.* Both of these plants become sticky if left too long before

During the first year, sainfoin rarely makes a great show but improves vastly in the second, and in the third will astonish the grower. If milled seed is used, it can be sown with the same implement that is used to sow clover-seed.

Sainfoin has been the salvation of many a farmer on the poor, thin, chalky lands of the south of England. There are two sorts, the common and the giant; the latter is the one usually sown on heavy soils, as although it does not hold out as many years as the common sort, its yield in hay and feed is much greater. The treatment of the crop is generally as follows: 3 and sometimes 3½ bushels are sown to the acre with a grain crop and harrowed in, taking care to cover the seed well — in fact, in Kent, we always used to put it in with a grain drill at 7 inches apart — the next summer, it should be mown for hay before the blossom is more than half expanded. The aftermath is good for all sorts of stock, and the best place in the world for weaning lambs, as they never scour on it. We have known it stand for 12 years, but it is generally, in the usual course of cropping, ploughed up for wheat in the 7th year, completing the rotation, and avoiding the too frequent repetition of the red-clover: thus — turnips, barley, clover, wheat, which is the ordinary shift, would become turnips, barley, sainfoin, down for 5 years, wheat — a most refreshing course for the land if it will bear sainfoin. I have an indistinct recollection of Jonas Webb telling me that he had succeeded with the giant sainfoin on a clay-farm some way off from the Babraham establishment. The seed must be new, or failure is certain. We fear however, that any attempt to grow it where *white* clover fails to take would be hopeless — *no* plant will grow without plant-food, and we fear that the gentleman who asks a question as to the probable success of sainfoin on his "terre sablonneuse très médiocre", which is most likely utterly limeless, will not find any plant to answer his requirements. The sheep's foot would work wonders, and until that is tried we see no hope for the "very moderate sandy soils."

Show-Sheep

As to the best method of getting up a lot of sheep for show, there is the choice between house and open-air feeding. Sheep are not so happy under cover as in the open air, and we have heard the opinion expressed again and again that an open-air life is the best even for show-sheep. Any judge can at once tell a shed-fed sheep by from his wool. Plenty of room is also a point, and many prizes have been won by sheep which have been allowed to run forward in front of their fellows and pick the primest clover, rape, and cabbage. Upon the artificial foods it is not necessary to dilate, except in so far as to say that sheep of this description should be allowed a plentiful supply of the best that money can purchase. A constant variety in natural foods, and a liberal quantity of the *best linseed cake and old beans* fairly indicate the food; but who can describe the many minor points as to early and late feeding, frequency of meals, and methods of tempting the unwilling appetite, and coaxing the animals to grow? These belong to the art of shepherding, and are of vital consequence. A master might as well try to take prizes without sheep as without a shepherd, and it would not be possible to commit all the store of knowledge possessed by a competent shepherd to paper. Neither possible nor yet desirable; and if it could be done, the written directions would not ensure the same success in other hands. First-rate shepherds are not so uncommon as they are difficult to find, because they are not given to changing their situations often. A pleasant feature of sheep-farming is that mutual regard of master and shepherd, both men appreciating each other's value. Training is carried on with some little affectation of secrecy, and much undertoned and almost whispered consultation. The attention is constant and the daily care extraordinary.

The trimming of show sheep is a matter of importance. There are those who object to trimming, but it is impossible to show sheep in the natural unkempt and rough state. It is really cruel to ask a breeder to exhibit his sheep in a great show, before ladies and gentlemen, without dressing them. What would a horse-breeder say to a regulation insisting that his hunter or his thoroughbred should appear ungroomed and rough, with long tail and uncombed mane? A sheep-breeder has similar feelings, and similar failings. Besides, the public like to see animals well turned out of hand, and even the pigs appear with their hair curled and oiled, and their skins blooming as if they had been immersed in a bath composed of toilet vinegar. Trimming may be overdone, or unfairly done, but to the legitimate use of the art there can be no objection. The methods vary with every breed. The Leicester appears, like the parson, all shaven and shorn. The Lincoln is smeared over with some mysterious unguent, which makes the hands feel very disagreeable if they are allowed to touch the fleece. The Cots wold comes out curly in coat, white, and redolent of soap-and-water. The Southdown appears as like a plum as a sheep can possibly be made, and bears evidence of the shears over his entire carcase. A very smug gentleman indeed is the Southdown when in his war-paint. Trimming is carried to the greatest perfection in the Down races.

***

Some curious mistakes found in a "Practical treatise on Agriculture":

"Sheep-dung injures the quality of barley, causing it to yield less starch." What would our English farmers, on the chalk-hills of the Eastern and South-eastern countries say to this? Almost all the best barley used for malting in the large malt-houses of Saffron-Walden, Ware, &c., for the use of the gigantic breweries of London, Burton-on-Trent, and elsewhere, is grown on the chalk, after turnips fed off by sheep, who pass day and night in folds there; and the writer continues: "Sheep-dung is suitable to all soils except calcareous ones!"

The Sussex breeders keep large flocks of ewes, sell all their wether lambs and full-mouth ewes to the upland graziers of Kent, Surrey, &c., who fatten them; they send their ewe-tegs out to keep on the grass-lands at so much a score for the winter, which fully accounts for the small size of the Sussex-downs, as they return half-starved. Our old friend, Rigden who kept 300 breeding ewes, never fattened a single sheep, except the superb dozen or so of 20 month's old wethers he used to send to the Smithfield Club show at Christmas, carrying off many a prize.

The old English weaned their lambs on August 12th, hence called Lammas-day; but if your ewes have lambed, as they ought to do, by the middle of April, they might be separated from their young by the middle of July, and got into good condition with *rape* by the 1st of September, when, if the ram is introduced to them, they would lamb at the end of January or the beginning of February. What pulled down the price of early lamb this last spring (1892) was the scores of mean little cats, weighing some 4lbs. the quarter, that were sent up to Montreal in March. Every little "cag-mag" butcher had one hanging up in what he is pleased to call his *market,* and a wretched sight it was. A lamb should weigh, if properly done by, from 32 to 40 lbs. of carcase at 12 or 13 weeks old, and should not be slaughtered before that age; then, if he and his dam have been well fed, peas not having been omitted in the lamb's ration, and the ewe having had a fair allowance of cake and oats, the lamb will be a credit to his feeder, as well as to the butcher who kills him and to the cook who dresses him, and the ewe, in a fortnight or three weeks from the time the lamb goes to market, will be ready to follow in its footsteps.

Some people, who ought to know better, I have seen look for teeth in the upper jaw of a sheep: of course they did not find any! Nothing so easy as to tell the age of sheep, up to four year's old. In the sheep districts of the South of England they are called, by the number of teeth, "two-tooth, four-tooth, six-tooth, and full-mouth sheep:" ungrammatical, but sufficiently descriptive. A weaned lamb, with us, becomes a *teg*, and a ewe that has lost some of her teeth from age is a *crone*.

Chapter Eight - Treatment of Lambs After Weaning

WE left our lambs taking entire care of themselves for the first time in their short lives. There will, of course, be a trifle of fuss and bother for a few days, but they will soon get accustomed to their deserted condition, feed away merrily, play about with each other in bands of three or four, and pass the noontide hours in shady spots in refreshing slumber. By the time the July heats have invaded the country, and the grass begin to show signs of failing to supply sufficient food for the increasing needs of the growing lambs, the earliest sown of the rape will be ready to receive them; so we will now consider how we shall make the best arrangement for consuming the rape in the most economical way, not forgetting that we have two things to attend to: first, the proper feeding and care of the lambs; and, secondly that the land shall get its share of the manure from the lambs equally distributed over the surface and, as soon as possible, be covered in by the plough.

In order that this system may be thoroughly carried out, it is necessary that the flock should be divided into two lots: those intended for pushing forwards for the block, i.e. wether-lambs, and the ewe-lambs that are meant to carry on the duty of breeders.

Some of my readers, many of them indeed, have never seen a flock of sheep folded on a piece of rape or turnips. A fold is an enclosure made by hurdles (see cut), which can be easily trans-ported from place to place, and can be set down continuously in rows imper-vious, if properly arranged, to sheep.

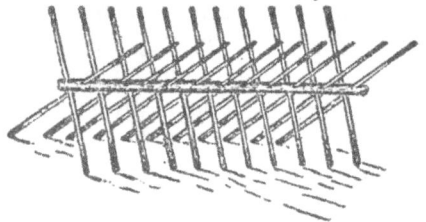

A 3-inch bar, 12 feet long, is pierced with holes alternated from side to side, and 6 inches apart, through each of which is driven any rough stuff in the form of stakes, 4 feet 6 inches long, exactly as in the engraving. As I mentioned before, when speaking, of another form of hurdle, if the sheep are given to jumping, a wire should be run through the tops of the stakes, so that, when a sheep tries to get out of the fold, he will find himself arrested by the wire, and after being thrown back by it once or twice, he will be quite cured of his erratic propensity. A lad of 15 years can easily roll over these hurdles, and set them in rows, so the work is not costly. Some arrangement, to prevent small, backward lambs from creeping out through the spaces — trifling enough — where one hurdle meets another, will have to be made: we used, when at Sorel, to drive short stakes into the ground, but a better plan would be to make, or have made, a dozen or so of iron tripods, as in frosty weather the stakes cannot be driven into the ground.

The number of these hurdles required on a farm will depend entirely on the number of sheep kept. As to the form in which the fold should be pitched, that again depends upon the shape and width of the field. At all events we must remember that the fold must not be too wide, because it is a maxim in folding that when one length of the field has been gone over by the flock, it should not be wider than a plough can turn over in a day's work.

But in the case we are considering, we must have two folds, and for this reason; we have two sets of lambs as we said above. The wether-lambs will naturally be given the freshest and best feed, that is, the first piece of rape or turnips enclosed by the fold, in which there will probably be a trough or two, with rations of chaff, or peas, or grain, and, towards the winter, all three; and when they have taken off the first bloom of the rape or turnips in the fold, another of the same size will be added at the end of the former, into which the wether-lambs will be introduced, and when this is occupied, the ewe-lambs will be allowed to follow the wethers in their partially over-eaten fold, to what is technically called "clean-up."

When the wether-lambs have finished their second fold of rape, etc., a third fold must be pitched a-head, into which they will be driven, and the ewe-lambs will again take their place. Next, the first fold-hurdles, being now not wanted in their original place, will be moved a-head of the fold now occupied by the wether-lambs, and so the whole ground-plan is completed.

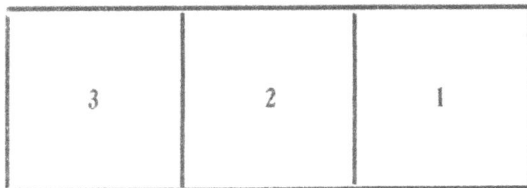

3	2	1

The Folds

I need not, I hope, insist upon the necessity of being very careful the first time sheep are put on rape, or clover. There is a great danger, if they are allowed to begin upon their new food on an empty belly, or when the rape, or clover, is wet with rain or dew, of their being we call "blown", and what the French call "météorisés", or "planet-struck", from one of those old, now wornout, superstitions, so many of which remain pertinaciously clinging to the less educated of our countrymen. The best way to ensure immunity from this danger, is to feed the lambs up well all the morning and so late as 3 P.M., and then to let them into their first fold for a couple of hours. The first thing, upon entering the fold, the lambs will do is to hunt about all over the place, treading down the rape, &c., and there is no chance of their beginning to feed until they have investigated every corner of the fold. Pursue the above plan for a couple of days, and after that they can be trusted to look after their own safety as far as gormandising goes.

Rape is but a watery food after all, though we have long been convinced by constant watching that the chemist's idea of the water in roots, and all green-crops is by no mean the idea of the practical feeder. But as rape is full of liquid, in some form or other, it is clear that the young sheep must have some kind of dry food to aid its digestion in the assimilation of the moister food; the question we have now to solve is: what dry food is most suitable to the delicate appetite of the, as yet, unaccustomed lambs?

Chaff, whether of straw or hay, is an excellent food — if one could only get the lambs to take it; the pity is, that until really cold weather comes on, the little wretches persist in neglecting it. Clover-hay chaff stands a better chance of being eaten, but even that is not to be depended upon. Of the grains, we prefer oats, and of pulse in the absence of our old friend, the horse-bean, we must resort to pease, and where can we find any superior in this line to the Canadian white-pea. It may not be generally known, but when a sample of "boiling pease" of English growth is offered for sale at Mark-Lane, the great English market for grain, pulse, etc., no corn-factor will offer a price for the bulk until a quart or so of it has been sent to one of the numerous taverns in the neighbourhood of the market and boiled as an experiment on its quality; while a sample of Canadian peas is bought at once, as if there were no doubt about their "melting in the pot", though we find here, that but too often they are refractory.

In my own practice, I always gave the dry food to the lambs in the afternoon, because, somehow or other, in this country, sheep do not seem to feed so heartily on rape and early turnips as they do in Britain; so an early feed or grain, etc., might take away their appetite for the green-meat. I do not think it necessary to crush the grain or pulse? because it is very rare to find any undigested grain in the dung of sheep, in fact they are capital chewers. As for the quantity of grain required for each lamb, I always found about a pint of pease and oats mixed, with a double-handful of chaff, sufficient; but towards the end of the season, as the mornings ' get colder and the lambs, or *tegs* as they would be called at home, are approaching the desired state of ripeness, to the pint of mixed grain and pulse I should add about half a pound of decorticated cottonseed-cake, and make the chaff of clover-hay alone. If the dung of the sheep was found to be too hard, half linseed cake should be substituted for half of the other kind.

The ewe-lambs, or future mothers of the flock, we left condemned to eat the leavings of their ungallant brothers; but in order to make up for this unkind treatment it would be only judicious to let them have some trough-food. Oat-straw chaffed, a trifle of cummins, otherwise maltculms, the rootlets of malted barley broken off after the kiln-drying, and sifted, and easily obtained in any of the towns of the province at a very low price, their value as food not being appreciated here. We ^w^ here the analysis of malt-sprouts as they call them in the States, as well as, for comparison, one of oats:

Oats.				Malt-sprouts		
Digestible nutrients.				Digestible nutrients.		
Protein	Carbohydrates	Fat		Protein	Carbohydrates	Fat
9.0	43.3	47		20.8	43.7	0.9

The relative value of these constituents per 100 lbs. of stuff is for oats $0.98, and for the malt-sprouts, $1.33, according to Wolf. You will observe that the malt-coombs have but little fat, but, *en revanche*, they contain more than twice the quantity of protein and albuminoids, the true generators of flesh or lean-meat; hence they must be good food for all young animals, who require plenty of firm-flesh-making food, and can do without much fat in it, if they find there plenty of starch to be converted into fat; starch you all know, of course, is a carbohydrate. Of the different recipes for feeding given in the books, we advise our readers to beware. Many of them may be locally beneficial and profitable, but too many are arranged without due consideration of cost, and too many are theoretically correct but fail utterly when practically applied.

If any one finds it more convenient to use corn than oats, he can try it, or when wheat or barley are cheaper per ton than oats, they might take the place of the other grain; but, as a rule, we have always found that oats and peas, mixed, in the proportion of one of peas to three of oats, make about the best food for young things.

I mentioned, just now, *horse-beans*; they do well here, if sown early and on heavy land. To be sown 30 inches apart in the row; as for the seed per acre, that depends upon the sized bean employed; of the small "pigeon-bean," the smallest, 2½ bushels an acre imperial will do, but the large "harrow bean" requires 3½ bushels — 1/6 less to the arpent. Beans should be got in, if possible, in April; harrow as they are coming through the ground, horse-hoe as often as convenient, and edge-hoe, i.e., along the rows on each side, taking the rows between the feet, once or twice. Many acres used to be grown on the Island of Montreal, whether their cultivation is as prevalent there as it used to be we do not know.

A good plan, as regards sheep-keep, is to sow, thinly, a little rape-seed between the rows of beans before the last hoeing. This will afford some useful feed after the beans are carried. As we said before, "old-beans" are one of the favorite foods for sheep in England. We have known them yield, in Essex, England, as much as 80 bushels an acre, and in Gloucestershire, on the alluvial soils of the Severnside, we have seen the haulm, or straw, nearly seven feet high. Very strong food is bean-straw for all kind of young-stock; better even than pea-straw: try it, you who farm heavy land.

If, at the beginning of September, you fancy your provision of late green-fodder for the sheep seems likely to run short, try a few acres of a mixture of Hungarian-grass and rape; say 3lbs of rape-seed and 15lbs of Hungarian-grass to the acre. The first frosts will turn the grass a little brown, but the rape and the grass together will do your sheep good, if you keep them out of it on frosty mornings.

When a length of the field has been fed off by the folded sheep, send the plough at it at once, and follow the same plan every time.

I was asked, some time ago, how late rape will keep good for sheep. My answer was, that I had kept fattening lambs, or tegs, on it till the 8th December, in 1884, and that they did as well on the last day as they had done in August. The land had been ploughed right up to the fold; the yield of oats, the next grain-crop, was 70 bushels to the imperial acre, nearly 60 to the *arpent.* The lambs or tegs turned out so well, that specimen-joints were sent to Montreal, among other persons to the late Mr. Wolferstan Thomas, President of the Merchant's Bank.

If properly managed, the tegs in the leading fold, that is, the wethers, should be all ready for the butcher by the 1st December, when they will probably be about 9 months old. Some of the more forward ones will most likely be ready to be drafted a fortnight before the above date. They will, we may well hope, present a somewhat different appearance to the miserable rats we so often see in the latter autumn months working their weary way along the streets of Montreal on their road to the abattoirs, with their long undocked tails hanging down nearly to the ground, and their thighs all daubed with the filth of the roads. Why should not a fair proportion of them be shipped at once to England? If the proper treatment had been pursued and the proper breed had been selected, these tegs should certainly weigh from 8 to 9 stones, i.e., from 64 to 72 pounds the four quarters, the very size required by the leading London butchers for their West-end trade. At the present time, Downs of that weight, in good order, fat, but not too fat, are worth, as we said above, at least 18 cts. a pound the carcase, that is, each sheep would sell from $11.50 to $13.00 a piece. Surely this must pay! [1]

The Ewe-Tegs.

Let us now suppose that the wether-tegs are all off the farm, and that we have only the ewe-tegs to care for. You wall have in store, at the beginning of the winter, a few turnips to begin upon, with swedes and mangels, clover-hay, pea-straw, oat-and-barley-straw: at least let us hope so. The turnips will not keep good long, and the swedes must then take their place; the mangels you will of course not touch — if you can help it — until the spring is near at hand. In the sheds, the older ewes will lodge with the ewe-tegs, as their food wall be the same, until the former are getting forward in lamb, when they must have an improved ration, to fortify them for the coming arduous task of parturition.

We were much impressed the other day by a passage in "Wrightson on Sheep management": "Turnips and straw are capable of keeping sheep in good condition, and so far as nutritive properties are concerned, they are sufficient. Too little regard is paid by purely scientific authorities to the great variation of quality in turnips according to the land upon which they are grown. To them a turnip is a root containing 92% of water, and this is enough to condemn it... White turnips, although actually inferior to swedes as a food, *according to analysis,* are superior to them up to January 1st, or even later. They are less trying to the digestion, and their consumption forms an excellent introduction to the harder winter-feeding which begins with the new year."

Curious language, is it not from the Principal of an Agricultural College? But, then, you see, Mr. Wrightson is a practical farmer as well as a man of considerable attainments in science. He probably knows that while the swede of the "plastic clay" formation and straw will only keep bullock in growing condition, the Aberdeenshire swedes and straw will make those admirable bullocks we used to see at the Christmas market in London. Our farm-tutor, Wm. Rigden, could not fatten sheep or bullocks with the swedes that grew on his farm at Hove, near Brighton, Bug., without cake and grain; while not 15 miles off, on the sandy land near Shoreham, at the foot of the South-downs, sheep were going up to London in scores every week, after receiving no other food than swedes and straw. What peculiarity there was in either soil no one knew, neither could chemical analysis discover. The Hove soil must have been of a great deal more than average fertility, for the wheat-crop yielded, one year with another, 48 bushels of the finest white-wheat to the acre.

As was mentioned before, ewes in-lamb need plenty of dry food. Pease-straw and clover-hay are the best "roughage", as our friends in the States call it. Water, of course, will be liberally afforded them. In early spring, when the tegs are losing their central teeth, the roots, whether swedes or mangels, must be cut for them. Prof. Wrightson cautions people about giving mangels to *wether* sheep. He might have included rams in his caution, as these roots are equally dangerous to all males of the ovine race.

Few things are fresher in my memory than the sad losses experienced by Wm. Rigden, whose name I have so often had occasion to mention. In 1852, he, as usual, was preparing a lot of his Southdowns for the annual show of the *Royal Agricultural Society*, then always held in the month of July. Though his farm was running over with green-meat of all kinds: rape, vetches, and trifolium (crimson clover), red-clover, and lucerne, he was paying 25 shillings a ton for mangels; as he told me, his pupil, he had always found his stock do better on that root than on any other green-food after the month of April.

But in this case, it turned out just the reverse, as regarded the rams, though the ewes did as well as usual. With a sad face, in walked the shepherd, one morning, while we were at breakfast: "Sir", said he, "I wish you would come and look at the sheds, there are two of the rams mortal bad, and I can't tell what ails them." Rigden and I went to the sheds at once, and there we found two of the best show-rams lying and kicking, evidently in great agony. A veterinary surgeon was sent for at once, and on his arrival, after he had examined the sheep carefully, all he could say was, that, from their actions the pains must be in the urinary apparatus. He prescribed certain remedies, which had no effect, and when the poor animals were clearly passed hopes of recovery they were slaughtered, and the parts indicated carefully examined. In the upper part of the passage, the urethra, was found, in each case a small crystallized lump, about as big as a hazel nut, that entirely prevented any emission of urine. Three more rams died in the same way, but not one ewe suffered in the least; and we need hardly say that, after this sad experience, no more male sheep ever tasted mangels on that farm.

Now, this is my experience in the use of mangels for male sheep; let us see what Mr. Wrightson says about it:

"My attention has lately been called to the fact that mangels cannot safely be given in quantity 'to wether-sheep. They appear to have an injurious effect on the bladder and other urinary organs, and in many parts of the country flock-masters dare not use them for wethers, although they may be given to female sheep. On this point of practice, I may say that mangel is relied upon as a change of food for ram-lambs in the great sheep-breeding districts of Wiltshire and Hampshire. The roots are either cut into troughs or thrown about among vetches or rape as a change, but the lambs have at that season so large a variety of food in the shape of green-fodder, that no evil consequences follow. Rambreeders would find it difficult to do without mangels in the hot months, and shepherds may be seen in our show yards slicing mangels and giving them with cabbage and rape to the rams".

Well; we present both opinions and facts to our readers, leaving them to take their choice. For myself I say, that I would on no account run the risk of losing valuable animals by supplying food to them that has been *known* to work such damage as that I have recounted.

Of timothy-hay I have the highest opinion as food for horses, but a low opinion of it as food for sheep; clover and sainfoin I prefer for the flock, particularly if a fair proportion of straw be cut up into chaff with it. Plenty of straw of all kinds should be always at the command of the sheep; given without stint in racks, it help to fill their bellies and therefore to keep them contented; we need hardly say that, after peas-straw, oat straw is the best and barley-straw the worst.

Of cakes, doubtless linseed cake is the best, then comes decorticated cotton-cake, and undecorticated last. But our favourite food has always been, for sheep in sheds, a mixture of peas, oats and linseed, or, perhaps more properly, flaxseed. We recommend the following mixture as having done its duty by us as long as we had stock to feed:

Take any quantity of the above in the following proportions:

Oats or corn	3 lbs.
Pease	2 "
Linseed	2 "
	7 "

Mix the above and grind them finely; then take a bushel of chaff, from straw of any kind, and mix the whole together, sprinkling the heap with a little water. This can then be distributed among the sheep in troughs and will be found most effective for fattening sheep, or lambing ewes, for fattening beasts or for milch-cows. I introduce oats into the mixture because if the flaxseed is not surrounded by something of the sort, the oil of the seed would ooze out from the stones in the act of grinding; the oatmeal and husks absorb the oil. In England, this kind of food was largely used as long ago as 1848, and here, where linseed can often be bought for from .70 cts to $1.00 a bushel it would pay to use it. I used, formerly, a small roller-mill that just cracked the seed, so nothing was wasted. Both rollers of the same diameter: like a good malt crusher, in fact. We often see in the papers advice about using linseed and skim-milk for calves. The writers recommend steeping the seed in water for several hours, or boiling it. If any one of my readers will take the trouble to take a grain of linseed into his mouth and try to crack it, he will find that before long a thick coat of mucilage will surround it and render it impossible for him to masticate the seed. The experiment of giving unbruised linseed to cattle was tried many years ago in England, and it was found that, out of a thousand grains thus administered, eight hundred pass through the animals undigested.

No animals require such constant supervision as sheep; they are always getting into trouble: "At one time, the farmer, on his rounds, finds one sheep on its back in a furrow; a fat sheep on its back, or as the Scotch call it, "lying awald," is before long a dead sheep; or perhaps a lamb is entangled in the rough stuff in a corner of the fence; something is always happening, so that in order to obviate these accidents, the shepherd on the extensive sheep farms of Wiltshire and the Wolds of Lincolnshire, and on the Borders of Scotland and England, always have near the fold a small hut on wheels, wherein is stowed all sorts of medicines, salves, &c., likely to be needed, and in this hut the shepherd takes his meals, which are sent to him by his wife. The hut is, of course, moved from place to place wherever it may be wanted.

Sheep should be given fresh food daily, if possible. Here, it would be difficult to manage it, but, at all events, shift the fold as often possible. Three days' consumption at a time is the greatest space permissible. Mr. Wrightson objects to the use of the hurdle already depicted and with reason. The trouble must be enormous, if the folds have to be advanced half a dozen times a day, as often, that is, as the sheep need more food. When two folds are occupied, as I recommend there is no fear of much food being wasted.

[1] We are, and not unfrequently, blamed for repeating ourselves, and we must own that there is a great deal of truth in the accusation. But what is one to do? As a great writer said, in reply to a similar charge: "If an important principle is laid down only once, it is apt to be unnoticed or forgotten by the majority of readers. If it is inculcated in several places, quick-witted persons think that the writer harps too much on one string."

Chapter Nine - Shearing Sheep

A very interesting question arises at once on our approaching the subject of shearing the flock: Should the wool be washed on the sheep's back, that is, before shearing, or after it has been shorn?

Washing sheep. There has been a good deal of discussion in England lately on the question whether sheep should be washed before shearing or not; but as the opposition to the invariable practice of British flock-masters came from a suspicious quarter, I do not think that shepherds will blunt their shears over many more fleeces of unwashed wool than heretofore. The interest of certain of the manufacturers and staplers was the only thing regarded; the health of the sheep and the profit of the farmer were not considered at all.

However, the real state of the case may be gathered from the opinions of certain persons well skilled in the trade, who have no purpose to serve in decrying this invaluable practice of washing sheep before taking the fleece off.

Of two agents, buyers for large Yorkshire firms, the former speaks as follows: the value of wool not washed would be considerably less than the value of washed wool; also, unwashed wool would lose its colour much sooner than that which has been washed previous to shearing, and, consequently would not answer the purpose for which it was required, and, if kept, would considerably deteriorate in value.

The other agent writes: We could not buy any unwashed wool unless we could get it for six-pence or seven-pence a pound, [1] Do not be led away by what the papers say about not washing the sheep; it will be a great loss to growers who do not wash.

A large wool-merchant writes to an inquirer: I beg to say that the difference between washed and unwashed wool is from 25% to 35%. For instance: washed wool at, say, from 6d., 8d., 10., 12. a pound, would be for unwashed, 4½ d., 6d., yd., 8½ a pound.

Mr. Turner, another man well known as an expert in wool, and who in consequence of his reputation as such was chosen as judge of fleeces at the last meeting of the Suffolk Sheep Society (England), spoke strongly in favour of washing before shearing; and gave the following reasons for his opinion:

"Mr. Turner said he had taken great pains to go into this important question, and he had been asked over a hundred times by letter what was his opinion on the subject. He had no hesitation in saying that he was very strongly of opinion that English wool ought to be washed on the sheep's back. Objection, of course, was taken to this in many cases, but if they knew as much as he about the source from which these objections came they would not think much about them. In 1887, 23,000.000 lbs weight — about one-sixth of the whole produce — was exported from England, and if the wool was not washed this export trade would be ruined. For instance; there was an import duty on wool to the United-States of 5d. a pound; whether the wool was washed or not the duty was the same. It was quite clear that under the circumstances the Americans would buy washed wool which would only shrink 20 per cent., when they certainly would not buy unwashed wool which would shrink 40 per cent. The Bradford Chamber of Agriculture had made an inquiry into the subject, and he believes their decision would be against the non-washing of sheep."

If dirty unwashed wool is offered for sale, immediately the buyer begins to find fault, and to make out that there never was such a filthy lot brought to his factory; whereas, when a nice well-washed fleece is shown, the bright, clear look of the wool fascinates the eye of the manufacturer, and he jumps at the lot in a moment.

No! I do not think we will give up washing our sheep yet.

The Bradford Chamber of Commerce, the largest company of wool-buyers in the world, have recorded their opinion on the very point at issue. "What is wanted," they said, "is more care and attention to this part of his business on the part of the British agriculturist; and as the out spoken opinion of this Chamber may be of service, the committee appointed to report on the subject venture to say that in this matter an amount of culpable slovenliness prevails on the part of the farmer which in any other branch of our national industry would not be tolerated." If the Bradford chamber of Commerce have seen reason to alter their opinion since they reported to Earl Cathcart in the above terms, some few years ago, I shall be glad to have intimation of it.

Sheep Shearing

It may be taken as an axiom in economics, that the more completely finished for the use of the consumer any article is, when it leaves the manufacturer, the higher proportionate price will it fetch. For instance; cotton yarn is much dearer in proportion than the rough cotton as the bale leaves the press in its native country; and it is clear, from this consideration, that the labour expended on the cleansing, teasing, and other manipulations it undergoes, with a proper addition to the price of the article, goes on accumulating, until at last the purchaser of a printed calico dress pays for the whole.

Thus, we have often wondered why the farmers of this province are so fond of shearing their sheep in the unwashed state. To begin with, it will be said that the wool is washed afterwards: true enough, but shearing a sheep with a dirty skin makes rough work, and moreover, wool washed off the sheep's back is deprived of its *yolk*, when dry feels harsh, and is in an unfit state for certain processes of manufacture.

In the year 1862 we superintended the washing of 60 sheep for the late M. Amable Demers, of Chambly. The affair was very simply managed: the sheep were penned in a temporary fold, by the side of the "petite rivière de Chambly;" a large tub was kept full of water, into which each sheep was plunged and thoroughly washed, the dissolved *yolk* acting as a soap; and after ten days, passed in a clean pasture, the sheep were shorn, so much to the satisfaction of the proprietor and the manufacturer (Mr. Thomas Willett), that the former presented me with a two year old fat wether in acknowledgement of my assistance. I say, that the sheep were kept in a *clean* pasture, because it is well that there should be no roads or earth-banks for them to soil themselves against. The practice of tub-washings, as distinguished from pool-washing, has long been in use in Yorkshire, England, and was the invention of Raspail, a French chemist, who observed that "when the wool is washed, this soap (yolk) is dissolved, and takes the salts with it. Hence it follows, that the water that has been used in this process becomes, at each repetition, better adapted for the purpose." Stephens, in his "Book of the Farm," objects to the practice, but he seems never to have tried it, and as a set off to his opposition, I think the fact that, in England, tub-washed wool always brings from a half-penny to a penny a pound more than pool-washed wool will be sufficient. At least it used to when I knew the trade.

Sheep should not be washed until the water has attained a temperature from 56°F. to 60°F. After washing, they should wait ten days or so, before shearing, as the wool must not only be thoroughly dry, but the yolk, the natural oil of the wool, must return into it again, and the new wool should have risen from the skin, before the old is taken off. Disregard to this particular renders shearing difficult, and certainly injures the appearance of the fleece. Generally speaking, one may wash the first week in June and shear in the second: if the water of small streams be used, it will be found warm enough by that time.

"The yolk being a true soap, soluble in water", says Luccock, "it is easy to account for the comparative ease with which the sheep that have the natural proportion of it are washed in a running stream." The composition of yolk was found to be^ in the rough: soap of potash, carbonate of potash, acetate of potash, muriate of potash, lime, and an animal fatty matter which imparts to wool its peculiar odour. The medium quantity of yolk in short-woolled sheep, according to Youatt, is about ½ the fleece. More yolk is found on the breast and neck of the sheep than on any other part of the body, and it is there that the finest and softest wool grows. Softness of the pile is, therefore, evidently connected with the presence and quality of yolk. There is no doubt that this substance is designed to nourish the wool and to give it richness and pliability. In what way is the growth of the wool promoted? By paying more attention than our farmers are accustomed to give to the quantity and quality of this substance possessed by the animals which they select for breeding purposes. The quantity and. quality of the yolk, on which farmers seldom bestow a thought, and the nature of which they neither understand nor care about, will, at some future period, be regarded as the very essential and cardinal points of the sheep — considered as a wool-bearing animal. I must add to Mr.

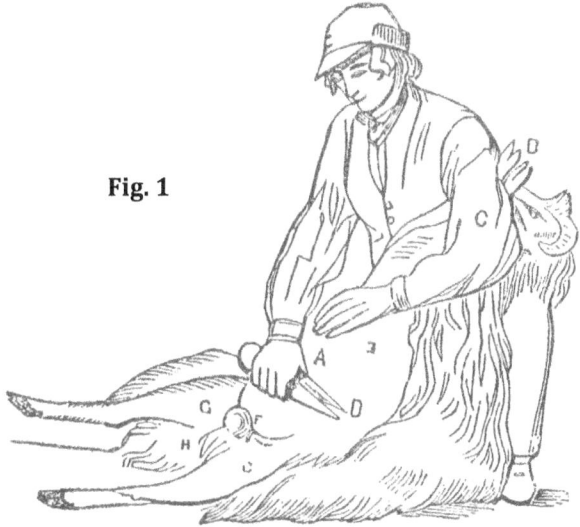

Fig. 1

Youatt's expression of opinion: for wool is so low in price and mutton so dear, that the question nowadays is: which sheep will produce the greatest weight and finest quality of meat?

Shearing. — A smooth barn-floor is the best place for this operation. Our flocks are so small that no extensive preparation is needed. The best shears have additional springs between the handles to separate the blades more forcibly, but they hurt the hand, and are not worth the trou-

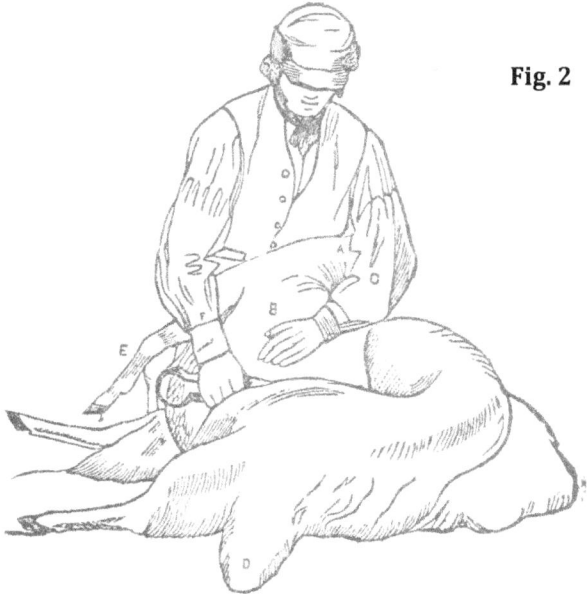

Fig. 2

ble. The great thing in shearing is to keep the points clear of the skin by gently pressing the blades upon the skin — keep the hand low, and rest the broad part of the blades upon the skin — you will not cut your sheep much if this is attended to. With scissors, such as we have seen used in the French country, but ragged work can be made.

Our engravings illustrate the three stages of shearing; First, after setting the sheep on its rump, and on the supposition that the clipper is a right-handed man, he rests on his right knee, and leans the back of the sheep against his left leg *a*, bent. Taking the shears in his right hand, and holding up the sheep's mouth with his left, he first clips the short wool on the front of the neck, and then passes down the throat and breast between the fore-legs to the belly. Then, placing the fore-legs *b* under his left arm *c*, he shears the belly across from side to side down to the groins. In passing down the belly and groin, where the skin is naturally loose, while the shears *d* are at work, the palm of the left hand *e* pulls the skin tight. The scrotum *f* is then bared, then the inside of the thighs *g g*, and lastly, the sides of the tail *h*. These are all the parts that are reached in this position. For the clipping of these parts small shears suffice; and as the wool there, is short, and of a detached character, it is best clipped by the *points* of the shears, as carefully held close, like *d*.

Fig. 2 represents the *second* stage of clipping. Its position for the sheep is gained by first relieving its fore-legs *b* from their position in fig. 1, and, gently turning the sheep upon its far side, while the shearer, resting on both knees, supports its far shoulder upon his lap. You may always rely upon this fact — the more a sheep feels at ease, the more readily it will lie quiet to be clipped. Supporting its head with his left hand, the clipper first removes the wool from behind the head, then around the entire back of the neck to the shoulder-top. He then slips its head and neck *a* under his left arm *g*, and thus having the left hand at liberty, he keeps the skin tight with it, while he clips the wool with the right, from where the clipping in the first position, fig. 1, was left off to the backbone, all the way down the near side. In the figure, the

fleece appears to be removed about half way down the carcass; the left hand *b* lying flat, keeping the skin tight; while the right hand *e* holds the shears at the right part, and in the proper position. The clipper thus proceeds to the thigh and the rump and the tail *d*, which he entirely bares at this time.

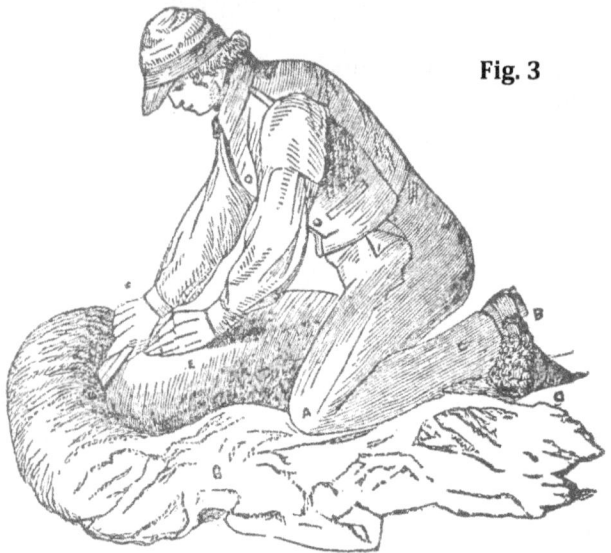

Fig. 3

Clearing the sheet of the loose parts of the fleece, the clipper, holding by the head, lays over the sheep on its clipped or near side, while still continuing on his knees; and he then rests his right knee, fig. 3, over its neck on the ground, and his right foot *b* on its toes, the ankle keeping the sheep's head down to the ground. This is the *third* position in clipping. The wool having been bared to the shoulder in the second position, the clipper has now nothing to do but to commence where it was then left off, and to clear the fleece from the far side from the back-bone, where it was left off in fig. 2, in the second position, towards the belly, where the clipping was left off in the first position, fig. 1. — the left hand *e* being still at liberty to keep the skin tight, while the right hand *f* uses the shears across the whole side to the tail. The fleece *g* is now quite freed from the sheep. In assisting the sheep to rise, care must be taken that its feet are free from entanglement with the fleece, otherwise, in its eagerness to escape from the unusual treatment it has just received, it will tear the fleece to pieces. [2]

On comparing the attitudes of the clipper and of the sheep in the different stages of clipping just described, with those of a mode very common in the country, it is necessary to look again at the *first* stage of the process, fig. 1, the common practice of conducting which is to place the sheep upright on its tail, and the clipper to stand on his feet, supporting its back against his legs — which is both an insecure and painful position for the sheep, and an irksome one for the man, who has to bow much down to clip the lower part of the animal.

In the *second* stage, fig. 2, the man still remains on his feet, and the sheep upon its rump, while he secures its head between his legs, in order to tighten the skin of the near side, which is bent outward by his knees. The skin is certainly tightened, but at the expense of the personal ease of the animal; for the hand can tighten the skin as well, as shown in all the figures, at B and E; whilst the bowing down so low, and so long, until he clips the entire side, can not fail to pain the back of the clipper. The third position is nearly the same in both plans, with the difference in the common one, which keeps the left leg bent, resting on its foot — a much more irksome position than kneeling on both knees.

[1] Best down teg wool was then worth from 12d. to 13d. a pound in England.
[2] The artist has erroneously represented the sheep lying on its *far* side, and the clipping to proceed from the belly to the back-bone, which is the proper posture for the second position, as also the keeping the head of the sheep down with the left leg *a*, whereas the sheep should have lain upon its *near* side, the wool been shorn from the back-bone to the belly, and the head *d* kept down with the right leg, as described above.

Chapter Nine - The Age of Sheep

As we said before, the age of a sheep up its fourth year, can be accurately told by inspecting its teeth. The following cuts and description we borrow from the book of the celebrated French Veterinary M. Samson.

A Two-Tooth Sheep, 12 to 15 Months

When a month old, the lamb has 8 temporary teeth and three molars on each side of the jaw.

At three months, a permanent molar is added to the three.

At nine months, the second permanent molar makes its appearance.

At fourteen months, two permanent incisors appear.

At eighteen months, a third permanent molar shows itself.

A Four-Tooth, 20 to 24 Months

At twenty-one months, there will be seen four permanent incisors.

A Six Tooth, 30 to 36 Months

At thirty months, there will be seen six permanent incisors.

And when the sheep reaches the age of from forty to forty-five months, he will show his *full-mouth*, that is eight permanent incisors.

A Full-Mouth, 40 to 45 Months

We must observe however, that many variations in the dentition of sheep occur, owing to the food, or to some other cause; but as a general rule, we may depend upon a sheep with two permanent incisors being one year old; if he has four he is two years old; if six, three years old; if eight, he is a full-mouthed sheep, i.e., four years old.

Note. — Cutting-up a sheep. Before being slaughtered, a sheep should be fasted for at least 24 hours. A sheep is supposed to lose about 13 per cent of its weight during the above-named time. Simple enough, the operation of slaughtering: the animal is laid on its side on a bench or stool; its legs, all but the hind leg that is uppermost as it lies, are tied together; and its life is taken by a thrust of a straight knife through the neck, severing the carotid artery and the jugular vein of both sides; the blood flowing freely out, the sheep soon dies. Many butchers, for what reason I know not, twist the poor thing's neck round, and break it against their knee. As soon as the sheep is dead, a small hole is cut through the skin into the flesh just where the leg begins to thicken; the butcher, applying his mouth to the hole, fills the carcase with wind on that side, and does the same on the other side, all the time thumping the carcase with his fist, to promote the equal circulation of the wind. This is said to obviate all risks of bruises.

Henry Stephens, in his great work, "The Farm", says that "sheep require no fastenings with cords," wherein, in our opinion, he his wrong. About 80 years ago, our farm-steward — *bailiff* we call that officer in Kent — was killing a sheep; the animal was not tied; the man, wanting to change the sheep's position, put the knife between his legs; the sheep in its struggles kicked the knife, it struck into the man's thigh, cut the femoral artery, and in less than ten minutes the man had bled to death. Since that time, no one in that part of the world has ever killed a sheep without tying its legs. One leg, the upper one as it lies, is left loose in order that the animal by his kicking with it, may hasten the expulsion of the blood.

The entrails are then removed by an incision along the belly, after the carcase has been hung up by a bar placed through the tendons of the leg or hocks. The parts inside the brisket and neck are now carefully washed out with a cloth dipped in water, the caul separated from the paunch, and the job is complete; though I forgot to say that the belly is distended with a large skewer or stretcher.

The carcase should hang for 24 hours in a clean, cool, dry room before it is cut up. If the room is not cool, meat will never become firm; if it is not dry, the carcase will remain clammy to the touch. In England, our sheep used to be divided down the back with a chopper; here, we believe, it is done with a saw.

Now, how shall we cut up the sheep into joints? That depends upon whether the sheep is a small Welsh sheep or a fair sized Down or Leicester. In the former case, the Scotch plan of making what, after the French system of butchering, is called a "jigot" (cf. gigot), i.e., cutting part of the loin in with the leg. On the English plan, the leg is cut off short and the loin is taken clean away from the neck. The shoulder is removed entire and either roasted — never boiled — or dressed with a fine savory sauce, after being egg-and bread-crumbed and carefully broiled on a rather slow fire, or in a Dutch-oven before the fire.

The best joint in a sheep's carcase is the saddle. It consists of the two loins undivided. If you are fortunate enough to have the means of roasting meat by an open fire, do not allow the spit to be thrust through any joint, but get what is called a "cradle-spit", in which the meat lies like a patient baby in its real cradle. You probably like mutton-chops; try one cut from the upper end of the saddle right across the back. This, commonly called a "double-chop", though it is by no means common, is as much better than a chop from the loin as a saddle is better than the loin itself; probably, from the gravy being kept in, instead of oozing-out during the progress of the broiling. Chops should be cut about 1½ inch thick, and broiled quickly on a sharp, brisk fire. Seven minutes should cook a chop of the above thickness.

The shoulder, plainly roasted, is eaten with onion sauce. The neck has two parts, the "best-end", which can be either roasted or boiled; if the latter, caper-sauce should accompany it; the "scrag" makes good mutton-broth. The breast, or brisket, is generally too fat. but is not bad when dressed something in the same way as I have described for dressing the shoulder.

Not many days ago, my butcher, who is good enough to think I know something about mutton, sent us the rib of Cotswold sheep. It was preposterously fat, but the tenderness and rich flavour of the lean was something to dream of.

Chapter Ten - The Diseases of Sheep

HAVING cursorily gone over the general points of interest in the treatment of the flock, I now turn, in the last place, to the diseases that afflict it; and, first, I will offer a few words on that troublesome complaint.

This is a complaint with which we have not much acquaintance in this country; still, it may be as well to learn something about it. It is generally prevalent in wet seasons, and sometimes affects hares and rabbits as well as sheep. In 1838, when I was staying at Wenvoe Castle, Glamorganshire, it was fearfully destructive to these; I remember well picking up five or six hares, in one short stroll through the meadows, all of them carried off by rot. Not many years ago, the writer's brother, of Hill-Court, Gloucestershire, wrote to me, saying that, on his property, there had been neither hare, rabbit, not sheep spared for nearly four years! The losses of sheep in England amounted, during the period, to millions, and no remedy has ever been found for the disease. Mr. Bakewell, the Leicester man, used to utilise this complaint in a curious way.

Having observed that some meadows enjoyed (?) the curious property of always rotting the sheep that were pastured therein, he made a point, whenever he wished to fatten off a lot quickly, of sending them to feed there; would any non-observant man have ever discovered that rotten sheep fatted more rapidly than sheep free from the disease? Of course these were not breeding sheep, but sheep intended for immediate slaughter.

The Rot

I am not veterinary surgeon, so I do not pretend to write an original treatise on this trouble. The disease is caused by the incursion into the liver of what is usually known as the *flounder* or *fluke*, the history of whose life is rather peculiar. It starts as a parasite of a fresh-water snail, whose shell is from one-fourth to one-half of an inch long; a well-known veterinary surgeon tells us that this snail suffers as much from the invasion of the fluke as the sheep does, and often dies from its attacks. "A severe winter kills off "rotten snails," but a mild winter may simply render them torpid, and thus unusually early cases of rot in sheep frequently occur. In the tissues of this snail it develops into a kind of sac, in which develop numerous larvae with tails. These escape through a special vent in the sac, and for the time lead a free and independent life in water. They then come to rest, cast their tails, and develop an enveloping cyst (bladder) of a snow-white colour which adheres to the leaves or stalks of grasses or water-plants. These may remain a few weeks, but if they undergo no further change of condition, the embryo within perishes. In this form, on grass or in water, they pass into the alimentary canal of the host or ultimate bearer (the sheep or horse), the pupa-cyst (the envelope) is digested by the action of the gastric juice, and, from the stomach of the host, through the duodenum and the bile-duct, the fluke gradually winds its way up to the liver."

It is hardly worth while to enter into a discussion on the different opinions of learned men as to the way in which the fluke is disseminated. Suffice it to say, that as one sheep may contain a thousand flukes, and each fluke forty thousand eggs, it is clear that one sheep may contaminate a large pasture.

When the eggs that the fluke has laid are expelled from the sheep, the rain washes them into drains, ponds, &c., the eggs develop into ciliated (hairy) embryos that swim about freely in the water, and after a few days of free, active existence, the young creature, losing its cilia becomes a creeping larva, and finds its way into the mollusc (snail, here), its first host as the sheep is its last.

Four stages of "rot" are:

1. Acute inflammation of the liver.
2. Dropsy; the liver is pale and firm.
3. Liver gets, so to speak, worn away.
4. Flukes, generally in May and June, leave the liver.

As we mentioned in a former part of this essay, sheep, in the earlier stage of the invasion of the fluke, profit in some mysterious way and absolutely get fat very quickly; in consequence of which peculiarity, butchers sometimes put their sheep on unsound land shortly before slaughtering.

We never heard of any sure cure for the rot; keeping sheep out of low-lying, unchained land is the best preventive. After severe, frosty winters, after a series of dry summers and falls, the disease is less prevalent than when those seasons have been wet. Says Prof. Wrightson: "To cure a rotten sheep is looked upon by many as hopeless. If, however, the animal outlives the expulsion of the flukes, it may recover. Professor Simond's recipe will prove a good guide, and is as follows: "Take of bruised linseed-cake and peas meal of each a bushel; 4lbs. each of salt and aniseed; 1lb. of sulphate of iron; grind all these together finely, and give to each sheep from half to one pint daily. Salt is an excellent tonic, and no sheep grazed on salt-marshes are ever affected by the rot. Change of diet is also insisted upon, and stimulants, such as turpentine and sulphuric ether, are recommended. Fortify the system in every way possible, by good hygiene, tonics and stimulants." Above all things, keep your sheep out of dangerous pastures; such as are known in your neighbourhood as "always giving sheep the rot."

Thrush or Canker

Every one who has kept breeding ewes knows this complaint. It affects both lamb and dam, though in different ways. The lambs suffer from sore mouth, the ewes from pimples and a sort of scab on the teat. The first sign of its appearance is that the ewes will not stand still to be sucked; then, naturally, the lambs wander about hungry and dissatisfied, continually trying to get food and continually getting discontented. Cure there is none; it has to be borne with, and only a change to warm weather and plenty of green-meat have any alleviating power over it.

When ewes are troubled with sore udders, the symptom of this complaint, they should be freely lanced, the lambs being of course taken from them and "brought up by hand." A mild dose of Epsom salts, with a trifle of coriander seed or aniseed should be the first remedy, followed by a little powdered columba root and salt in their trough-food twice daily. As a dressing for the udder, a 5% solution of carbolic acid, applied twice a day, will aid in healing wounds and dispersing the eruption — a weak solution of alum will do as well. The mouths of the lambs should be washed out with a solution of chlorate of potash, 10 grs. to the oz., just enough to moisten the surface without allowing any to be swallowed. Where proud flesh appears on the gums or lips, it should be carefully touched with lunar caustic, and any loose tooth should be removed. Youatt recommends that the mouth should be washed two or three times a week with a solution of alum, or diluted tincture of myrrh, and that a couple of oz. of Epsom-salts should be administered.

No doubt, the disease is caused by damp lodgings, cold, raw weather, &c. Some assign it to feeding among harsh stubble, or stony ground; but though individual cases may spring from these causes, we cannot think they will account for the wholesale losses experienced by some of our leading English flockmasters.

Foot-Rot

Of this malady, the treatment, as far as I know, is, or at any rate was, the universal treatment pursued by all South of England shepherds. Of "butyr" of antimony, which Youatt recommends, as I did in my previous recipe, he says: "There is no application comparable to this. It is effectual as a superficial caustic; and it so readily combines with the fluids belonging to the part to which it is applied, that it quickly becomes diluted and comparatively powerless, and is incapable of producing any deep or corroding mischief." If, in operating, the hoof has been too severely deprived of its horny covering, bandage it. Trim the feet of whole flock to prevent the wall of the hoof from growing too long so as to double under the foot. When paring the foot for the disease, previous to the dressing with butter of antimony, take care to get to the very bottom of the loose horn. I used to harden my heart when about this job, and it was needed, for the struggles of so gentle animal as the sheep in pain are by no means agreable to one who is not used to performing vivisection.

Fortunately for our farmers in this province, sheep are rarely affected by that horrible complaint, the foot-rot. Still, some years ago, it would seem, from the reports of the Chicago Market, some recently imported Shropshires were attacked with it shortly after their arrival. The buyer made a terrible fuss about it, and no wonder. The seller, whom I know to have been utterly ignorant of the mishap, was called all sorts of name, as if he could have detected the latent malady. I never could understand why long-wools should be so much more subject to this complaint in the feet than Downs and other short-wools; but such is doubtless the case; a very nasty complaint it is, and nothing but patience and persistent attention will cure it; contagious too, without a doubt, and, I believe, worse in wet than in dry seasons.

The foot-rot attacks the hoof, in the division, and gradually works its way upwards under the horn. If sheep are kept on damp straw in the latter autumn and in spring, that will be a favourable opportunity for the foot-rot to "get in its work." As I mentioned before, there is not much danger in allowing the bedding; to accumulate in hard weather; but in mild seasons, I should prefer, if there is any chance of foot-rot affecting the flock, making a floor of boards, laid, say, three-quarters of an inch apart and about 3 feet from the ground. These boards should be swept twice a day, and occasionally sprinkled with some disinfectant. The droppings could be drawn out from under the boards, and, mixed with bone-dust or superphosphate, would be a famous manure for rape or turnips. Do you fancy the sheep would not like to repose on boards? just watch them in summer, and you will see that they always select the hardest parts of the field to lie on in their noon siesta. To preserve the urine, a few bushels of spent tan-bark might be used in the pit under the boards as an absorbent, and if the disinfectant is frequently used, there will be no perceptible smell, and the nitrogen will of course be preserved intact.

But the question we were considering is, how to cure the foot-rot. Well, I have cured it with my own hands and although it takes time and, by no means pleasant, trouble, I do not think any one need despair of success, if he will follow these instructions to the letter. We are almost certain to have the foot-rot here, sooner or later, so that it is right that we should be prepared to face it when it comes.

With a sharp knife and a steady hand, pare away all the horny part of the hoof that has been undermined, taking care not to cut into the flesh so as to make the blood flow. Then, apply with a feather, all over the part left bare, some *butter of antimony*. Mr. Stephens calls this cruel treatment, but the disease is far more painful than the remedy, and that pain is continuous. The flesh will smoke under this treatment, but if you will apply it earnestly the animal will be cured, and, at any rate, it is preferable to letting the poor thing die in pain, as it certainly would were it to be allowed to go on long untreated. This, as far as my experience goes, is the only cure for the foot-rot. There are several remedies prescribed in the books, such as driving the flock through newly slaked lime, &c., but I never knew them do much good.

Flies

There are three sorts of flies that are injurious to sheep: the common fly, the tick-fly, and the botfly.

Of the common fly there are several kinds, of which the house-fly or *musca domestica* is one. Whether this unwelcome and persistent worrier of the human race ever gets into the country and persecute the sheep, is a matter of doubt; but there is no doubt that several of his brothers, the *diptera*, or two winged flies, exercise that malevolent office. "Buzzing round their tails, settling on the dung, and busily looking over for a place where she may lay her young, are various forms of those larger and even more offensive creatures, the meat-fly or blue-bottle, or bronze-coloured cousins of the same kind."

"These creatures, which are the horror of the pedestrian, the horse-man, and the shepherd, are well known to all of us; they frequent country sides near the bush, wherever wooded districts are found, but do not care so much for open pastures."

Of the sheep, two vulnerable points are their favourites, the head and the tail. Unfortunately, the Southdown breed, having less wool on their heads than other sheep, are not so well protected from the attacks of the fly as are the Shropshires, the Cotswolds, and, in a measure, the Hampshire-downs. As soon as one fly lights on a trifling sore spot, whether caused by the friction of a piece of dirt, or what not, he sets to work to make further discoveries; soon, others join in the work; they crowd on to the poor thing's unhappy head and dive into the flesh for a juicy suck, until a broad scald extending all over the forehead is effected. You may easily tell when the fly is at work: the sheep runs from place to place, with his head down; stops suddenly, stamps his feet, in fact is nearly crazy with irritation. How can a sheep thrive that is in such torment? How comfortable do you feel when the "black-fly" attacks you as you are salmon fishing below Quebec?

Hogg, the "Ettrick Shepherd", gives a simple remedy against the fly. "I happened to be assisting at the sorting of sheep of the Cheviot breed, where sundry of their heads were broke with flies, [1] The shepherds brought them out of the fold with the intention of smearing the sore parts with tar. I advised them to anoint them with coarse whale-oil, such as they mixed with the tar, having several times seen sores softened and healed by it. Some of it being near at hand, they consented. The flies were at this time settled upon the fold in such numbers, that when we went among the sheep we could with difficulty see each other, but to our utter astonishment, when those who were anointed were turned among the rest, in less than a minute not a fly was visible!

Whether this application of whale-oil was a new discovery or not, I do not know; but it is certain that whale-oil, commonly called train-oil, is now looked upon as a sure safeguard against the attacks of the common fly. A little sulphur mixed with the oil is an improvement. As it is so invariable in its beneficial effects, it is a matter of astonishment that it is not more generally known.

Of the sheep cap I spoke already at length but it must be borne in mind, as I said before, that a cap must never be put on a head that is already sore, lest an attack from the fly should have produced eggs, and the maggots from the eggs should have hatched and begun their ravages without the shepherd having had a chance to espy them. Some maggots are bred in the dung or in the dirt adhering to the coat, and soon make their way into the flesh. There, they produce redness, soreness and an exudation of matter. The symptoms exhibited by the sheep are pretty much the same as those mentioned above.

The mineral known as "mercury-stone," is said, by Mr. Wrightson, to quickly kill the maggot, so shepherds should always carry a piece of this with them. It is just like a piece of slate pencil, and can be got at any chemist's. To part the wool, and thoroughly rub the place affected with the mercury-stone, speedily kills the maggots, which ought, however, to be brushed out by the hand until they are all got rid of. Spirits of tar also has a good effect, and my good friend, Mr, Henry Gray, of St. Lawrence Main Street, Montreal, has a very good mixture for the purpose.

When maggots are neglected, the sheep pulls his wool off with his teeth, and, later, the skin becomes hard and dry and comes off just as if it had been burnt or scalded.

The Sheep Bot-Fly

This fly is of the same order as the horse-fly. Miss Ormerod, the great authority in England on entomology of all kinds, describes it as being rather larger than the common house-fly, and of an ashy colour, spotted with black. Youatt says "that if only one of these pests appears, the whole flock is in the greatest agitation." His vile aim is to lay his eggs just in the inner margin of the nostrils: they soon hatch, and at once crawl up the nose until they reach the recesses of the frontal sinuses, where they hang by means of their tentacles till the following summer.

During the passage up the nostrils they are very troublesome, until they reach their peculiarly selected home, when they do not seem to do much harm. Again, when they are ready to find an exit by the nose, in order to change into pupae, the constant sneezing of the sheep is an evidence that they are suffering. The life of the bot-fly is decidedly interesting, but as our authorities give no cure for its attacks, we will not pursue it any further.

The Gid or Sturdy

Almost every sheep-breeder has noticed that from time to time, not frequently, thank gracious, a mysterious complaint attacks lambs, generally with fatal effect. This is the *sturdy* or *gid*, so called because the lambs under its influence are invariably giddy and turn round from right to left, or the reverse; sometimes standing as if fixed to the spot, and refusing to move on. Gid is another form of animal parasitism, and owes its origin to one or more cystic, i.e., bladder-like parasite in the brain.

This parasite invades the brain in the following manner: the bladder-worm, as it is sometimes called, varies in size from a hempseed, when young, to a cricket-ball, when full grown. They are derived from a tapeworm which infest the dog, and are in reality the young of that creature. Just as the maggots that we find in flesh are the young of the blow-fly, so are these hydatids (water-filled sacs) the young of the tapeworm. Each hydatid, however, represents not one but a brood of young tapeworms. How many, may be known by counting the little white spots on the walls of the bladder, which are, in fact, the heads from which future tapeworms are to spring, should they, by any chance, find their way into the belly of the dog. Without the dog, and other members of the same species, there could be no gid-parasites, and, consequently, no giddy sheep. The lamb acts as a host towards the bladder-worm, that is, it affords it protection, shelter, and food, until it is fitted for its new home in the dog, where it begins at once to throw out segments and become a complete tapeworm. It is all very wonderful; without sheep, and others of the same species, there could be no tapeworm. Neither the hydatid nor the tapeworm can be pleasant guests; the former by its growth and expansion breaks down the brain and paralyses its host; the latter reduces the dog to a skeleton, and sometimes causes its sudden death by epilepsy.

To get rid of the gid, it must be remembered that its habitat is the brain, and that, consequently, great care in the manipulation is necessary. Some puncture the bladder or cyst with a sharp instrument, and draw out the fluid with a syringe; but as this generally requires to be repeated, it is not much to be depended upon for a cure. All the dogs on the farm should be watched closely, and if any segments of the tapeworm are seen in their excrements, they should be shut up and not allowed to leave the kennel until the worm is completely got rid of. A dose of areca-nut, followed by a good dose of castor-oil, should be administered, and every atom of tapeworm, expelled by the medicine, should be burned.

The Tick

When sheep are seen to take every opportunity of rubbing themselves against posts, &c., it may generally be concluded that they are troubled with the *tick*. The little beast is small, with a red head, and its body is of a pale yellow-colour. The tick frequents both sides of the sheep's neck, and the inside of both shoulders and thighs. When sheep are dipped, ticks will not often be found troublesome.

Diarrhoea

Commonly called by shepherds, *white scours*. Not dangerous if taken in time. Our usual cure for it, in the South of England, was a change on to a piece of sainfoin. *Sain-foin*, from the French signifies "wholesome hay." Nothing causes scouring so often as a change from a bare pasture to a very luxuriant one; or a sudden flush of grass after a spell of rain, will have the same effect. A table-spoonful of castor oil, with a dash of ginger in it will do good but attention to the food of the flock is the main thing. Too much tampering with physic in the herd or the flock is a dangerous thing, and now that veterinary surgeons are to be found in every the smallest town in the province, one of these practitioners should be called in if any unusual symptoms appear to invade the flock.

A Heroic Treatment

(From the French)

"*An experiment.* In July, 1885, I tried the following experiment. Having found in the fields one of my Shropshire ewes, with a lamb at her foot, that was very weak and poor in condition, hardly able to stand, and at the point of death, I had her, and her lamb, taken at once to the sheep-shed, about three miles from the place where I found them.

"She was very ill; her nostrils were choked up, and her head so obstructed that it seemed as if suffocation must immediately take place. The following is the treatment I pursued:

"I drenched her with three half-pints of linseed tea, and, half an hour after, I poured into the ears three half-pints of warm brine from the pork-barrel. The next day the dose was repeated; within thirty minutes after this treatment, she was up on her feet and trying to eat some hay. For food, I gave her a little bran with some oatmeal and hay; her drink consisted of three half-pints of linseed tea thrice a day. After three days of this treatment, I gave her the lamb again, which she suckled as before: and in a week's time she was feeding away as usual in the fields. The lamb I sold in September for ten dollars, and the ewe remained with me in perfect health."

As for making ewe-milk cheese, we have seen quite enough of that on the "Borders." The pail has to be placed under the tail of the ewe when she is being milked, and, in spite of all the care taken by the milker, it is easy to see that filth, both liquid and solid, will fall into it: not a nice idea! Mr. Chapais states that the Larzac ewes, from whose milk the celebrated Roquefort cheese is made, give, after having weaned their lambs, 55 quarts of milk, from which 24 lbs of cheese are made, in 160 days, which is equal to 2.40 oz. a day. Would the value of this pay for the labour expended? Would it be worth while to wear out the ewe, who [needs all possible means of recovering her strength after weaning her lamb, for such a trifle as 24 pounds of cheese?

Mr. Chapais makes the milk of the ewe contain seven and a half per cent, of butter and six per cent of casein. Stewart, in his book, "Feeding animals," gives the following as the average composition of ewe's milk:

Casein	4.50
Butter	4.20
	8.70

Whereas, according to Mr. Chapais, the ewe of Larzac gives:

Casein	6.00
Butter	7.50
	13.50

Leaving a balance in favour of the Larzac ewe of 4.80!

[1] The essayist is not answerable for Mr. Hogg's English.

www.ingramcontent.com/pod-product-compliance
Lightning Source LLC
Chambersburg PA
CBHW022127280326
41933CB00007B/574